Angelo Marinucci

Theoretical Principles of Relational Biology

Space, Time, Organization

Springer

Angelo Marinucci
Philosophy
Universidade Estadual de Campinas
Campinas, São Paulo, Brazil

ISSN 2661-8915 ISSN 2661-8923 (electronic)
Human Perspectives in Health Sciences and Technology
ISBN 978-3-031-39373-0 ISBN 978-3-031-39374-7 (eBook)
https://doi.org/10.1007/978-3-031-39374-7

This Springer imprint is published by the registered company Springer Nature Switzerland AG
The registered company address is: Gewerbestrasse 11, 6330 Cham, Switzerland

Human Perspectives in Health Sciences and Technology

Volume 6

The Human Perspectives in Health Sciences and Technology series publishes volumes that delve into the coevolution between technology, life sciences, and health sciences. The distinctive mark of the series is a focus on the human, as a subject and object of research. The series provides an editorial forum to present both scientists' cutting-edge proposals in health sciences that are able to deeply impact our human biological, emotional and social lives and environments, and thought-provoking theoretical reflections by philosophers and scientists alike on how those scientific achievements affect not only our lives, but also the way we understand and conceptualize how we produce knowledge and advance science, so contributing to refine the image of ourselves as human knowing subjects and active participants in a constantly evolving environment. The series addresses ethical issues in a unique way, i.e. an ethics seen not as an external limitation on science, but as internal to scientific practice itself; as well as an ethics characterized by a positive attitude towards science, trusting the history of science and the resources that, in science, may be promoted in order to orient science itself towards the common good for the future. This is a unique series suitable for an interdisciplinary audience, ranging from philosophers to ethicists, from bio-technologists to epidemiologists as well to public health policy makers.

In nova fert animus mutatas dicere formas corpora (Ovidio 2008)

Dedicated to hopefulmonsters

Preface

A scientific turn is often a matter of a changing perspective, as a new way of looking at the same phenomena or area of knowledge from a new view point. This changing perspective may also be based on the choice of a different priority: same objects or conceptual structure but different epistemic evaluation of "what comes first". In a sense, the transition from Pythagoras ontological priority of numbers to Euclid's geometry, based on axioms to "producing/trace," segments, lines, circles, parallel lines etc., establishes different priorities: spatial symmetries precede the numbers.[1] We all know what happened, two thousand years later, when some decided to look at the planetary systems from a different perspective, that is from the point of view of the Sun.

This book is based on an approach to life that does not look at individual organisms first, as biological objects, but gives epistemic priority to "the spaces of biological relations". An ecosystem is not seen as the result of interacting organisms, but organisms are analyzed as the result of ecosystemic relations. A tissue is not seen as a network of interacting cells, but cells are understood as components or the result of a tissue structure. What matters first is not what cells are, individually, but what they do, collectively, as a tissue network. Of course, this does not forbid to work on cells and on organisms, but always in a context, indeed in a historical context: an organism or an ecosystem is the result of a history and nothing can be understood of them without a historical perspective.[2]

The book makes extensive references to physics, its history, its philosophy. As a matter of fact, the author first worked in those areas (Marinucci et al. 2021). The references to physics allow to understand the methodological analogies and the theoretical differences. Typically, we learned from Einstein's General Relativity Theory the priority given to interactions over the abstract, "*a priori*" (or even

[1] Euclid's axioms maximize symmetries and proofs are based on rotations and translations, i.e., symmetries of the plane.

[2] An organism can only be defined and analyzed as the result of a phylogenetic history (Lecointre et al. 2001); but also the first questions a doctor must ask you concern your age and clinical history.

absolute, Newtonian) space structure. To make it short, for GRT « the geometry of relativistic spaces is a tissue of interactions: when deforming these interactions, the tissue and its geometry change; conversely, a deformation of the geometry changes the interactions, their tissue » (Bailly and Longo 2006). This relativistic approach to space geometry and the role of interactions is an explicit source of inspiration for the method driving the book. That is, this perspective, inspired from physics, is assumed by the author as a method to construct biological theories (for ontogenesis and phylogenesis), while it allows to depart radically from existing physical theories. The latter are grounded on invariant properties first, by which objects and their trajectories are defined in physical spaces, as geodetics in suitable "phase spaces" (spaces of pertinent observables and parameters).[3] For the author instead, variation comes first and this in a space of changing relations, where it is impossible to fix *a priori* the pertinent observables and parameters. This may still allow to analyze individual organisms in an ecosystem or individual cells in a tissue, but after or only as a component of the intended relational structure. The Tissue Organization Field Theory (Sonnenschein and Soto 2007) is a paradigmatic example for the author.

Of course, this change of perspective has a long history. First Darwin, who proposed the two principles of heredity that allow to understand biological historicity: reproduction with variation and selection. Then the organicist perspective. The issue is to relate and enrich these two theoretical frames, by overcoming, in particular, some limits of the organicist approach, as stressed by the author. By the latter I mean a long tradition which goes back to Waddington and MacKlintock's epigenetics to Maturana and Varela's autopoiesis (Moreno and Mossio 2015), without forgetting their old philosophical antecedents: Kant and Bergson, in their diversity. And here comes a delicate issue. The autopoietic approach has been recently revised by the work by Montévil and Mossio (2015): the "closure of constraints" is an in depth analysis of how processes produce constraints that canalize/enable processes; this has been further extended to a "relational approach" in Montévil and Matteo (2020). By the introduction of (characteristic) time and an opening to historicity (I insist on my preferred motto: "nothing in biology can be understood if not in the light of evolution" or, more, generally, "...in the light of time"), Montévil and Mossio first set the grounds for a link of an enriched autopoiesis to evolutionary change in novel terms. According to the author of this book, though, the stress on invariant structure of processes and constraints (their closure), though "relative and historical", does not allow to move to the radical "relational historicity" of life: in the book's epistemic view, first a relational space changing in time, then the temporary stability (historicized invariants) of closures of constraints, at the core of organisms' organization. In the author's words, the closure of constraints is viewed as a "historical empirical *a priori*" in relational spaces. The debate is open, starting from, I believe, the same anti-reductionist school of

[3] Einstein first baptized his approach "Invariantentheorie". As H. Weyl stressed: "all fundamental principles in physics are based on symmetry properties"—symmetries, as groups of transformations, are the mathematical frame for invariance preserving transformations.

thought, yet with a further deepening of the relational perspective that may have an epistemological and scientific relevance: what comes first, conceptually? where to act first in experiments and knowledge construction? The example of the etiology of cancer (origin, prevention, and therapies), extensively mentioned in the book, is a paradigmatic case.

The book introduces several relevant new notions for biology, such as the notion of "thickness", as relating what is "possible" in a contingent context to the "real" enabled by it. An evolutionary niche enables a biological dynamics, while being changed by it; as the author stresses, enablement acts on "possibilities", as it makes "real" a possible process, a new organism for example. Thickness is a tool for the analysis of this passage on the grounds of existing organisms and phenotypes.

The book, by its extended references to physics, beautifully reinforces a strong anti-reductionist approach: it is the understanding of the rich history of physics and its theoretical inventions, such as the courage to radically change theory on the grounds of a change of scale (e.g., quantum mechanics vs classical or relativistic theory), which helps the author to drive away from physical theories. Historicity and relationality of life are at the core of the approach, a strongly needed insight in a time where the ecosystemic relations are misunderstood by a mechanistic approach to life. This contributes to misunderstand (and disrupt, by GMOs, endocrine disruptors. . .) the historical fine tuning of species and organisms, a relational structures, when approached by the genocentric/programmable views of life.

Paris, France Giuseppe Longo

Acknowledgments

I would like to thank Giuseppe Longo, Ana Soto, and Carlos Sonnenschein, for their leading about biology and mathematics, as well as Francesco Marinucci, Vanni Zavarella, Fabrizio Pignatelli, Luca Del Monaco, and Érika Barbosa Pereira for their support.

Introduction

A falling object is a generic falling object; in other words, it is not necessary to establish its specificity in order to define how a certain law (in this case "gravity") applies to it. Hence, once any falling object has been described, the dynamics of all falling objects are known.

On the contrary, a sick living being is *not* a *generic* sick living being because the course of its disease depends, for example, on any allergies to those substances, which normally do not cause reactions in most living beings, at least of the same species. In any case, it is always necessary to consider its singularity, its unavoidable contingency, despite the fact that evolution has strongly canalized at least the organisms within the same species and their functions. Once the reactions of a certain living being to a certain medicine are known, the reactions of *all* other living beings are *not* known, at least in the same species. From this point of view, *similarities* among members of the same species can be considered either as an expression of a fundamental invariant or as elements that underline their diversity.

Obviously, the one proposed is just an example, but it is sufficient to put an irreducible difference between physical object and the biological one. The fact that physical object is generic and its trajectories are specific (necessary), while biological object is specific (singularity) and its trajectories are generic (possible) (Bailly and Longo 2006) will be shown.

Since the late 1970s, theoretical approaches have been proposed that sought to complexify deterministic and reductionist biology of Monod and Jacob, using probability[4] as the central element of biology. This approach, like the organicist one (Varela 1979; Rosen 2005; Montévil and Mossio 2015) putting organic structures at the center of theory, offers extremely interesting investigation tools, but they still remains within a theoretical biology, conceived starting from physics.

The idea of this text arises from the necessity to build a new theoretical and epistemological basis for biology. Although biology has made enormous progress in the last seventy years, many biologists are still tied to the approach of Monod

[4] In this regard, Kupiec's works are particularly noteworthy (see bibliography).

and Jacob, even if (almost) no one claims that ontogenesis results in a "revelation" (Monod 1989, p. 87) of information contained in DNA or that there is a junk DNA. However, the few improvements in theoretical framework do not correspond to great advances in biology. Very often, it is just a matter of *ad hoc* adjustments. The attempt to actualize Monod's perspective, stating that information is not totally contained in DNA (Deutsch 2012), is an example, because its method and its epistemological grid, used to interpret life, are not questioned. In this sense, relational biology re-discusses and overcomes epistemological principles and tools of molecular biology: DNA conceived as a code, the metaphor of program, and an epistemology constrained within the narrow limits of logic and discrete mathematics.[5] In fact, there is a huge difference between rejecting the notion of "junk DNA" and building a theoretical perspective in which such statement are no longer possible.

By rejecting an epistemology dominated by logic, relational biology recovers *geometric tradition* of Euclid, Newton and, above all, of Poincaré and symmetries theory in order to avoid falling into the determinism and reductionism typical of molecular biology.[6] It allows to look at the "shape" of objects, without assuming that they are reducible to their constituents.

From the theoretical point of view, if all approaches, proposed since the 1960s of the past century, present a biology based on physics, with fundamental invariant, generic object, and specific (necessary) ontophylogenetic trajectories, relational approach intends to construct a biology different from physics, even if not in contradiction with the latter. In this sense, as mentioned, biological object is specific and the trajectories are generic (possible), and moreover, relational biology is based on variation: its *fundamental invariant is variation*, the real engine of life.

Generally speaking, biological object or observable is constituted relationally; in this sense, it will be proposed that *theoretical observable* of biology is *relational space*, manifesting itself in various forms: tissue, organ, organism, etc.

The *acritical* use of physical theoretical model in biology[7] is essentially due to the lack of a theoretical inspection about the differences between various scientific disciplines, jumping from physics to biology without analyzing whether this transition is possible.

Normally, the same mathematical and theoretical tools are acritically used in classical physics, quantum mechanics, and biology. The lack of an epistemological inspection leads to confusions between these scientific disciplines. It has to be

[5] Obviously, various approaches, such as probabilistic and organicist, will also be discussed in this book, which shift theoretical attention to elements that molecular biology leaves in background, such as structures and probability. These are certainly extremely important contributions, but insufficient to constitute a general theoretical framework because they still conceive biology from physics.

[6] As regards reductionism, see Monod (1989, pp. 88–89). As regards determinism, it is sufficient to mention the use of discrete mathematics and stereospecificity. These topics will be discussed and criticized in the first chapter.

[7] For example, in the Introduction to *Chance and necessity*, Monod talks about a "physical theory of evolution" and in Monod (1989, p. 104) he states: "the fundamental biological invariant is DNA".

recognized that contradictions between classical and quantum physics have produced a debate that continues to this day. For this reason, it is necessary to clarify and delimit the domains of aforementioned three sciences in order to define their limits and subject matter. This theme will be introduced in the first part of the book and definitively treated in the second one, because it is important to introduce new philosophical-biological concepts, aimed at clarifying the domain of biology. Generally speaking, it is necessary to schematically present these distinctions from now on.

As for classical physics and quantum mechanics, it is possible to say that if the former describes its objects *directly* and *maintains* a symmetry between what is calculated and what is measured, the latter describes its objects only *indirectly* and does *not* maintain a symmetry between what is calculated and what is measured (Bailly and Longo 2006). Describing an object directly means that calculations represent the object that can be measured. On the contrary, quantum mechanics describes its object indirectly, in the sense that it directly deals only with probability function of possible states of its object. Furthermore, as Schrödinger's cat mind experiment shows, what is calculated (the states of the cat) is not what is measured. In this sense, if Rovelli (1996, 2010) states that quantum mechanics describes *how* objects interact, it is possible to affirm that the two physics answer different questions. Hermeneutically, if classical physics answers the question "what is the object?", quantum physics answers the question "how is the object?".

What said implies that in classical physics everything that is possible is able to enter into domain of reality. On the contrary, in quantum mechanics, as Schrödinger's cat shows, not everything that is possible is able to enter into domain of reality. For this reason and in order to avoid confusion, it is necessary to distinguish the domains of the two physics:

Classical Physics

$$\text{Realm of reality}\begin{cases}\text{Possible or potential reality}\\ \text{Actual reality } (\textit{hic et nunc})\end{cases}$$

Quantum Physics

$$\text{Realm of possibility}\left\{\text{Quantum possibility (quantum coherence)}\right.$$

$$\text{Realm of reality}\begin{cases}\text{Possible or potential reality}\\ \text{Actual reality } (\textit{hic et nunc})\end{cases}$$

If all possibilities of classical physics are able to enter into domain of reality, then they can be considered "possible realities", i.e., possibilities *within* the domain of reality. Hence, it is possible to underline the clear distinction between the domain of possibility and the domain of reality, typical of quantum mechanics (Heisenberg 1971), in which possibility and possible reality are not the same as in classical physics. In this sense, it is even clearer that the object of classical physics is *real*, while that of quantum physics is *possible*, in the sense that it is a matter of *probabilities* (not referring directly to real object). In fact, it is possible to enter into reality when quantum coherence is lost.

In any case, all possibilities are given *a priori* in the two physics, while in biology this does not happen. This is one of the main differences between physics and biology. Moreover, as mentioned, if in physics objects are generic and trajectories are specific (necessary), in biology objects are specific (singularities) and trajectories are generic (possible) (Bailly and Longo 2006). It will be shown that this implies that it is necessary to consider biological object in its singularity, in its contingency, albeit canalized by its evolutionary history. Precisely the irrepressible biological contingency requires that the door to "radical variation" (hopefulmonster) is always kept open. It represents an ever-present, yet indeterminable possibility. From this point of view, biology deals not only with possibilities but also with the constitution of possibilities.

The scheme that will be proposed toward the end of the book is the following:

Relational Biology

Contingent possibility {
- Historical-empirical *a priori* (Evolution)
- *Enablement* (costitution of possibilities)
- Singularity of living being
- Thickness (Relation and evolution of possibilities)

Contingent reality {
- Potential reality
- Mesurement
- Actual reality

In this scheme, there are new concepts which will be exposed in the text and which cannot be explained now. For now, it is important to reiterate that there are strong and often unavoidable differences between classical physics, quantum mechanics, and biology.

This book is basically divided into two parts. The first one is devoted to relational space. The first chapter discusses the limits of Monod and Jacob's molecular biology[8] from both a theoretical and an epistemological point of view. The second chapter is devoted to the concept of relational space and analyzes the differences between physics and biology on the use of probability in random phenomena. The last chapter of the first part exposes the philosophical elements of relational biology, related to the concept of space. In particular, the concepts of singularity of living being, of constitution of possibilities and of contingency are shown. New concepts,

[8] This approach will be called "traditional" and "orthodox".

indispensable to this new approach, are also introduced, such as enablement (Longo et al. 2012) and thickness.

It is necessary to underline that the concepts of the first part will be taken up again in the second part and interpreted starting from the concept of biological time. In fact, the second part of this book deals with biological times and rhythms, interpreted from the relational point of view. After briefly dealing with the relationship between time and movement, the concept of entropy is discussed in physics and biology, with particular attention to "anti-entropy" (Bailly and Longo 2009; Longo and Montévil 2012). The second chapter of the second part takes up and defines some concepts of the first part such as enablement, thickness, causality, etc. After introducing the concept of organization, the last chapter of this book is devoted to evolution, limiting analysis to the evolutionary *interpretation* of organisms. A more detailed text on application of relational biology to evolution is a work in progress.

The general idea of this book is to set forth the theoretical basis of relational biology, which is necessary to interpret life newly. It is precisely to create a basis for combining the humanities and the life sciences, proposing a new model that also has concrete applications on the use of technologies and health sciences. In fact, the last part of the book deals, for example, precisely with *biological organization* and *historical time* and how to interpret them ontophylogenetically.

Bibliography

Bailly, Francis, and Giuseppe Longo. 2006. *Mathématiques et sciences de la nature*. Paris: Hermann.

Bailly, Francis, and Giuseppe Longo. 2009. Biological organization and antientropy. *Journal of Biological Systems*. 17th ser. 1: 63–96.

Deutsch, Jean. 2012. *Le gène. Un concept en évolution*. Paris: Seuil.

Heisenberg, Werner. 1971. *Physique et philosophie: la science moderne en révolution*. Paris: A. Michel.

Lecointre, G., H.L. Guyader, and D. Visset. 2001. *Classification phylogénétique du vivant*. Belin. https://books.google.com.br/books?id=tSAUAQAAIAAJ.

Longo, Giuseppe, and Maël Montévil. 2012. The Inert vs. the living state of matter: extended criticality, time geometry, anti-entropy – an overview. *Frontiers in Physiology* 3: 39. ISSN: 1664-042X. https://doi.org/10.3389/fphys.2012.00039. https://www.frontiersin.org/article/10.3389/fphys.2012.00039.

Longo, Giuseppe, Maël Montévil, and Stuart Kauffman. 2012. No entailing laws, but enablement in the evolution of biosphere. In *GECCO Companion '12*. New York: AMC.

Marinucci, Angelo, Stefano Salvia, and Luca Bellotti, eds. 2021. *Scienza e filosofia della complessità*. Roma: Carocci.

Monod, Jacques. 1989. *Le hasard et la nécessité*. Paris: France loisir.

Montévil, Maël, and Mossio Matteo. 2020. The identity of organisms in scientific practice: integrating historical and relational conceptions. *Frontiers in Physiology* 11. https://www.frontiersin.org/articles/10.3389/fphys.2020.00611.

Montévil, Maël, and Matteo Mossio. 2015. Biological organisation as closure of constraints. *Journal of Theoretical Biology* 372: 179–191.

Moreno, Alvaro, and Matteo Mossio. 2015. *Biological Autonomy*. New York: Springer.

Ovidio, Publio Nasone. 2008. *Metamorfosi*. Milano: Garzanti.

Rosen, Robert. 2005. *Life Itself: A Comprehensive Inquiry into the Nature, Origin, and Fabrication of Life*. New York: Columbia U.P.

Rovelli, Carlo. 1996. Relational quantum mechanics. *International Journal of Theoretical Physics* 8th ser. 35: 1637–1678.

Rovelli, Carlo. 2010. *Quantum Gravity*. Cambridge: Cambridge Universit Press.

Sonnenschein, Carlos, and Ana Soto. 2007. *The Society of Cells*. New York: Taylor & Francis.

Varela, Francisco. 1979. *Principles of Biological Utonomy*. New York: North Holland.

Contents

Part I
Biological Relational Space

Deterministic Biology

The aim of this chapter is to explain the starting point of relational biology. It is possible to understand the necessity of a new theoretical perspective about biology only making clear the weaknesses of molecular biology and, in particular, the limits of its intrinsic determinism and reductionism.

In this sense, the reasons why the Monod (1989) and Jacob's (1970) biology will be named deterministic or orthodox biology will be shown. In particular, it is deterministic and reductionist like the physics before Poincaré and Boltzmann. In this sense, it is possible to conceive their biological approach to nature starting from the physics based on double implication of determinism and predictability.

In order to understand the Monod and Jacob's *philosophical* and epistemological perspective, it is necessary to underline the difference between "reduction" and "reductionism".

Reduction and Reductionism

"Reduction" is a scientific *practice* consisting in decomposing a phenomenon in the simple elements that describe it. Reduction has merely an essential heuristic value. It is very helpful in science.

However, "reductionism" is not a scientific practice, but it is a philosophical *concept*. According to the latter, in order to know nature, it is necessary to *decompose* it in the constituent elements, that allow to describe it *completely*.

The fundamental difference between reduction and reductionism is the *ontological commitment* that reductionism implements.

In this sense, the switch from reduction to reductionism involves an essential and theoretical shift: the *method* is not anymore a path, useful to know some aspects of nature, but it becomes the *imposition of a rationality* to the nature itself. The history of thought is rich of these methodological shifts. Ptolemy said that the aim of an

A. Marinucci, *Theoretical Principles of Relational Biology*,
Human Perspectives in Health Sciences and Technology 6,
https://doi.org/10.1007/978-3-031-39374-7_1

astronomer is to show that celestial phenomena are uniform circular movements. Galilei said that:

> Philosophy is written in this grand book, the universe, which stands continually open to our gaze. But the book cannot be understood unless one first learns to comprehend the language and read the letters in which it is composed. It is written in the language of mathematics, and its characters are triangles, circles, and other geometric figures without which it is humanly impossible to understand a single word of it; without these, one wanders about in a dark labyrinth (Galilei 2005, pp. 237–238).

Fourier said that "mathematical analysis is as extensive as nature itself" (Fourier 2009, p. XXIII). Anyway, it is very important to underline this large ontological commitment concerning nature and mathematical instruments. This idea is based on the assumption that it is always possible to know the essence of nature and to discover its secrets and its constituent elements.

From Galilei to Fourier (but it could be also mentioned Laplace and Lagrange), there are several ways to give a sense and to radicalize the proposition: "mathematics is the language of nature". In any case, the gnoseological strategy is always the same in physics and in biology because, especially in orthodox biology, it is possible to find out the idea that mathematics is the language of nature. This idea is elucidated in biology by the concept of genetic program. However, in *deterministic* biology DNA is the fundamental element of living beings and it is "written" like a program. In this sense, it is necessary to decipher genetic code, in order to read the nature language. On the contrary, relational biology agrees generally with Wittgenstein, when he says:

> We predicate of the thing what lies in the method of representing it. Impressed by the possibility of a comparison, we think we are perceiving a state of affairs of the highest generality (Wittgenstein 1986, §104).

Generally speaking, the ontological commitment involves to link the object to well-determined properties, in order to consider it *liable* to scientific knowledge. Therefore, an object is liable to knowledge, if it is an object of a theory and if it respects some characteristics. In other words, an object is never considered in itself in a theory, but it is always a *constructed* object.

For example, if calculus is the language of nature, then a scientific knowledge is possible only if it is possible to apply it on natural phenomena.[1]

Instead, in relational biology, the instruments of knowledge allow only to establish "images" (Boltzmann 1905a) of nature and coherent structures by which to

[1] Israel (2003) suggests historical examples. It can be said that during the nineteenth century, scientists tried to apply the calculus to society. The problem is that the elements of population move and behave totally different from celestial bodies. For example, during the debate about the opportunity to inoculate the smallpox, D'Alembert proposed a contrary argumentation based on epistemological concepts. He was contrary to the use of probability in that mathematical argumentations, because probability is not trusted, since it has not a proper epistemological status. In general, it is possible to know the objects of a theory scientifically, if they respect criteria defining them as scientific objects. As regards traditional biology, just a fundamental invariant should allow to know life. In this sense, it should be possible a truthful scientific knowledge.

interpret it. After all, each theory allows to organize knowledge differently. Before explaining the gnoseological and methodological relation between the last concepts and the deterministic biology, it is necessary to dwell on reductionism.

Reductionism in Biology

As regards biology, it is important to clarify how Jacob and Monod conceive "reduction" and "reductionism", from the philosophical point of view, in order to delimit the possibilities and the intrinsic limits of epistemology that appear in *The logic of life* and in *Chance and necessity*. Monod states:

> But we must hasten to say that this "reduction to the microscopic" of morphogenetic phenomena does not as yet constitute a working theory of those phenomena. Rather, it simply sets forth the principle in whose terms such a theory would have to be formulated if it were to aspire to anything better than simple phenomenological description. This principle defines the objective to be reached but furnishes little in the way of clues of how to get there (Monod 1989, pp. 88–89).

From the philosophical point of view, Monod defends that *reduction* is not already a *theory*, but he considers the possibility of constitution of a *reductionist* theory because he talks about a "statement of principle": something stronger because this is at the basis of the construction of a theory. Monod focuses on this philosophical strategy: the methodological shift from the reduction to reductionism, the shift from the scientific practice to essence.

Monod's choice involves that, in order to know something, it is necessary to reduce the complicated or complex[2] phenomena to its simple elements, presuming that it is enough to combine them mechanically to explain natural phenomena. In the same reductionist framework, Jacob suggests the "machine" and "program"'s metaphor to conceive life.

> The aim of modern biology is to interpret the properties of the organism by the structure of its constituent molecules. In this sense, modern biology belongs to the new age of mechanism. The program is a model borrowed from electronic computers. It equates the genetic material of an egg with the magnetic tape of a computer (Jacob 1970, p. 9).

In Monod's and Jacob's work and in molecular biology tradition, there is a considerable use of the metaphor of "program" in order to conceive DNA.[3] Regarding its philosophical sense, Jacob expresses it exemplarily in the following passage:

> Each egg therefore contains its entire future: the stages of its development, the shape and the properties of the living being which will emerge. The organism thus becomes the realization of a program prescribed by its heredity (Jacob 1970, p. 2).

[2] In a reductionist framework, "complicated" and "complex" are synonyms.

[3] As regards the analysis of this metaphor and its limits, see Longo's works http://www.di.ens.fr/users/longo/download.html and Bailly and Longo (2006), Longo and Montévil (2014).

Kupiec summarizes the Monod and Jacob's ideas accurately: the central dogma "states that the flow of information can only pass from DNA to proteins and that a gene determines a protein entirely by the information it contains. DNA contains [...] all the information necessary to make an individual. They are coded and form a real genetic program. If we were able to decipher this information, we could literally read the individual on the DNA" (Kupiec and Sonigo 2000, p. 67).

The philosophical strategy is based on the reduction of living beings to its constituent elements (genes and information). Consequently, a real individual being (for example a cell or a dog) is just a possible combination of simple elements, which are the fundamental biological invariant by Monod.

More generally, the space of possibilities would only include *combinations*[4] of constituent elements. This reductionist idea involves also that the established space of possibilities can include *nothing else*: nothing else that is not pre-determinable combinatorially. It is very important to underline it, because the possibility is clearly subjected to well-fixed constraints[5] and because biological *interactions* are only conceived *combinatorially*; consequently, they have *no epistemological status*. In short, element interaction does not allow to know anything new.

It is necessary to underline that in Jacob's book, various mentions about *contingency* can be found, although he never fundamentally calls into question either biological reductionism or determinism.

In order to explain the gnoseological role of combinatory and contingency, it is necessary to distinguish two kinds of biological processes. On one hand there are the "linear" processes where the environment does not play any role, on the other hand, there are the "contingent" processes where the environment plays a role in the choice of possibilities. As regards the first case, François Jacob writes:

> The very nature of the genetic code prevents any deliberate change in program whether through its own action or as an effect of its environment. It prohibits any influence on the message by the products of its expression. The program does not learn from experience (Jacob 1970, p. 3).

The procedure "DNA - RNA - proteins" is deterministic and linear. According to Monod, stereospecifity is a fundamental concept (Kupiec and Sonigo 2000) that, from the philosophical point of view, aims to reinforce the image of mechanism through "a protein finds its place in the cell because it is directed by an instruction" (Kupiec and Sonigo 2000). In this sense, a linear and deterministic process is directed *univocally and necessarily*.

---

[4] Jacob (1970, p. 3): "The genetic program, indeed, is made up of a combination of essentially invariant elements. By its very structure, the message of heredity does not allow the slightest concerted intervention from without. [...]. The program does not learn from experience".

[5] Jacob (1970, p. 3): "A given message thus represents a particular selection among all the arrangements possible. It is a particular order among all those permitted by the combinative system of symbols. Information measurements the freedom of choice, and thus the improbability of the message".

Regarding the second kind of process, it is a bit more complicated because it is necessary to consider the environment and the contingency that it involves.

If the space of possibilities can only be filled by the combinatory of elements, in some cases the environment will choose a possibility among the others. Thus, environment can only act on possibilities *already given totally a priori* and not on their constitution. These interpretations highlight the *epistemic* status of environment and contingency in Monod and Jacob. In other words, biological randomness is not intrinsic to the theory.

This kind of contingency allows and justifies a massive use of probability in biology (like physics) as well as the complete superposition of the philosophical concept of randomness and probability (a mathematical tool).

In deterministic biology and in classical physics, scientists state that it is always possible to determine *a priori* the *full* list of possibilities in a random phenomenon and to attribute a probability to each element of the space of possibilities, whose sum is 1.[6]

Starting from this situation, contingency will only contribute to realize a possibility (for example a phenotype).

If contingency plays an epistemic role and if the space of possibilities is filled combinatorially, it is possible to introduce the philosophical distinction between space of essence and space of accident. The space of essence contains all the *necessary and sufficient* elements to fill the biological space of possibilities *completely*. The space of accident contains all the elements which can not act on constitution of possibilities, but act just on their choice.

For example, if the space of essence contains DNA and if it is considered *like a program*, each part of not coding DNA can be called "junk DNA".[7] Even though today nobody names those parts of DNA "junk", at least from the philosophical point of view, there are not many progresses. It is not sufficient to increase the space of essence, for example introducing new elements or moving the information beyond the DNA (Deutsch 2012), it would be better to give up the clear distinction between essential and accidental space radically: this is the first condition to conceive a different and new biology.

This distinction reminds of the one from Laplace between essential and accidental (or "strange") causes. He works in the same theoretical framework of Jacob and Monod's biology: a reductionist and deterministic one. In this sense, contingency can be considered only outside this theoretical framework. Referring to Jacob, it can be affirmed that the radical *variation* is beyond the theory in the orthodox biology.

---

[6] On the contrary, if it is impossible to know *a priori* the complete list of possibilities, it is necessary to introduce a different interpretation of the concept of "relation" and to distinguish aleatory (a concept) and probability (a mathematical tool).

[7] By the way, in this framework, determinism and reductionism are so strong as the junk DNA is not even considered in the accidental space.

In this biology, there is a certain kind of possible variation, but it is never a *radical* variation, because the first type of variation is always subsumed (and preserves) to the fundamental invariant of biology, i.e. DNA.

> The rigidity of the program thus varies according to the operations. Certain instructions are carried out literally. Others are expressed by capacities or potentialities. However, in the end the program itself determines its degree of flexibility and the range of possible variations (Jacob 1970, p. 10).

As regards radical variation, Monod and Jacob use several times the verb "to appear" (*survenir*). In *The logic of life*, for example, the possibility of a fundamental biological change (a DNA mutation) appears (*survient*) at the theoretical framework of biology, because it is impossible to conceive it within reductionist and deterministic biology.

> Changes in the chemical text appear [*surviennent*], not by modification of a previously chosen sequence, but blindly (Jacob 1970, p. 289).

The meaning of this quote brings back to the title of Monod's book *Chance and necessity*. In the traditional biology, chance and necessity are and remain two opposed concepts: in order to understand the nature scientifically, it is necessary to discover some invariant,[8] building a theory of fundamental invariants. This perspective leaves variation in background, despite it is one of the fundamental elements of life, even its fundamental element.

In a theory of invariants (insofar as deterministic or reductionist), there is no place to conceive the variation in terms of mutations: it appears (*survient*) in the biological processes, never changing the development of a biologic procedure (Jacob 1970). Although it is possible to change some DNA elements, the mechanism of copy remains always the same. Therefore, orthodox biology approach forgets that variation is one of the most important invariants of biology and life.

Orthodox biologists "can" set variation outside the space of essence, because it is not considered in itself but just when it modifies the "genetic message":

> Only those errors which cause a change in the genetic message itself - that is, mutations - can have important consequences for the species, for once they have appeared, they are faithfully copied in turn from generation to generation (Jacob 1970, p. 288).

In this quote, Jacob proposes a very interesting link among "mutations", "mistakes" and "faithful copy". It is a coherent link in the theoretical reductionist and

---

[8] The nucleic-acid message does not learn from experience" (Jacob 1970, p. 292). "For the basic strategy of science in the analysis of phenomena is the ferreting out of invariants. Every law of physics, for that matter like every mathematical development, specifies some invariant relation; science's fundamental statements are expressed as universal conservation principles" (Monod 1989, p. 100). "The universal components - the nucleotides on the one side, the amino acids on the other - are the logical equivalents of an alphabet in which the structure and consequently the specific associative functions of proteins are spelled out. In this alphabet can therefore be written all the diversity of structures and performances the biosphere contains. [...]. The fundamental biological invariant is DNA" (Monod 1989, p. 104).

deterministic framework. Indeed, orthodox biology can only conceive processes as always identical iterations (Buiatti and Longo 2013).

> It [bacterium] grows and lengthens [. . .]. Then it cleaves into two, producing two bacteria identical with each other and with the original one[9] (Jacob 1970, p. 268).

After showing the fundamental elements of orthodox biology and the structure of their relations, it is necessary to explain the foundation of its gnoseology.[10]

Physics Before Poincaré and Boltzmann

The philosophical framework of orthodox biology and the one of physical approach to nature before Poincaré and Boltzmann are the same. It is very important to highlight some aspects of the physics in eighteenth and nineteenth century, because it points out the epistemological problems concerning also orthodox biology, although biologists have been normally ignoring.

Boltzmann: The Images of Nature

In *Chance and necessity*, Monod states:

> No performed and complete structure preexisted anywhere; but the architectural plan for it was present in its very constituents. It can therefore come into being spontaneously and autonomously, without outside help and without the injection of additional information. The necessary information was present, but unexpressed, in the constituents. The epigenetic building of a structure is not a creation; it is a revelation (Monod 1989, p. 87).

This quote shows Monod's reductionism very well. Considering that he interprets the fundamental invariant of life (DNA) by the metaphor of informatic program and by the philosophical hypothesis that life is essentially discrete, it is possible to affirm that orthodox biology proposes an ontological argument. Consequently, the idea of a program does not give an image of reality, but it *is the* reality. If "discretization" has an ontological role, then it is possible to know the essence of life, therefore it is possible to consider the noise like an accidental element of biological description.

---

[9] Today, scientists know that the similarity among bacteria is 30–40%.

[10] "Statistical mechanics made it possible to interpret the average behavior of large populations of molecules. Genetic analysis, however, revealed that biological properties were not the result of statistical molecular events; but that, instead, they were based on the quality of some substances contained in the chromosomes. In contrast to the order of inanimate bodies, the order of living organisms could not be extracted from disorder" (Jacob 1970, pp. 249–250).

On the contrary, the ideas of program, discrete, continuous etc. are just "instruments" to generate images of nature, but not to reach its essence. In this sense, Boltzmann reminds that:

> The differential equations of mathematico-physical phenomenology are evidently nothing but rules for forming and combining numbers and geometrical concepts, and these in turn are nothing but mental pictures from which appearances can be predicted. Exactly the same holds for the conceptions of atomism, so that in this respect I cannot discern the least difference. In any case it seems to me that of a comprehensive area of fact we can never have a direct description but always only a mental picture. Therefore we must to say, with Ostwald, "do not form a picture", but merely "include in it as few arbitrary elements as possible" (Boltzmann 1905a, p. 42).
>
> We must not aspire to derive nature from our concepts, but must adapt the latter to the former. We must not conceive that everything can be arranged according to our categories or that there is such a thing as a most perfect arrangement: it will only ever be a variable one, merely adapted to current needs (Boltzmann 1905b, p. 166).

Thus, in physics and in biology, it is necessary to avoid any reductionist methodological shift. So, the epistemology of Boltzmann could allow to reconsider truly biological theoretical framework.

Before Poincaré: Noise and Determinism

As regards orthodox biology, other issues require reflection, for example, how mathematical and conceptual instruments are used. The interpretation of non-linear differential equations has a general naivety. It is well known that this kind of equation can generate "deterministic chaos": very little errors below the measurement can produce trajectories diverging exponentially.

The problem is that, if non-linear equations are used in a *discrete* theoretical framework (access to the measurement is exact), it becomes impossible to understand the fundamental importance of noise[11] and that order can emerge from disorder. As Monod explains, "the invention of differential equations [is] a means for defining change in terms of what remains unchanged" (Monod 1989, p. 100), but it is fundamental to ponder the use and the interpretation of mathematical instruments.

As regards possibilities and limits of orthodox biology, Monod and Jacob's conceptual elements show that their biology is generally close to the world-view of physicists before Poincaré and Boltzmann.

The topic, which brings biology closer to physics, is the *measurement problem*. The metaphor of program requires that: first, the access to the measurement is exact; second, noise has no epistemological status and, finally, a reductionist and ontological commitment is required. Before Poincaré, the noise had the same

[11] It is very important to stress that computers, used for simulations, are discrete-state machines.

epistemological meaning as in deterministic biology. It is necessary to seek the reason in the interpretation of the relation between non-linearity and linearity.

In the eighteenth and nineteenth century the relation among physics, mathematics and nature was so particular that it is worth deepening. It has been said that the calculus was the language of nature; so, it is necessary to outline its fundamental reasons.

Now, it is impossible to reconstruct the history of the science between the eighteenth and nineteenth century. It is enough to focus on the most important moments in order to outline a general framework (Marinucci 2011; Blay 1992). Let's ponder these Lagrange and Fourier quotations:

> I intend to reduce the theory of this Science [mechanics], and the art of solving problems relating to it, to general formulae, the simple development of which provides all the equations necessary for the solution of each problem. I hope that the manner in which I have tried to attain this object will leave nothing to be desired. [...] No figures will be found in this work. The methods that I explain require neither geometrical, nor mechanical constructions or reasoning, but only algebraic operations in accordance with regular and uniform procedure. Those who love Analysis will see with pleasure that Mechanics has become a branch of it, and will be grateful to me for having thus extended its domain (Lagrange 1788, pp. v–vi).

This quote shows the aim to translate *Principia*'s geometry in the language of calculus. The approach of Lagrange to nature is clear: he reduces mechanics to a "field" of calculus. Therefore, physical problems coincide with their mathematical formulation.

In this period, the lack of the concept of mathematical "model" legitimizes to conceive a coincidence between the mathematics and nature. On his part Fourier affirms:

> The differential equations of the propagation of heat express the most general conditions, and reduce the physical questions to problems of pure analysis, and this is the proper object of theory (Fourier 2009, p. 6). [...].
>
> After having established these differential equations their integrals must be obtained; this process consists in passing from a common expression to a particular solution subject to all the given conditions. This difficult investigation requires a special analysis founded on new theorems, whose object we could not in this place make - known. The method which is derived from them leaves nothing vague and indeterminate in the solutions, it leads them up to the final numerical applications, a necessary condition of every investigation, without which we should only arrive at useless transformations (Fourier 2009, pp. 6–7). [...].
>
> mathematical analysis is as extensive as nature itself; it defines all perceptible relations, measures times, spaces, forces, temperatures [...]. Its chief attribute is clearness; it has no marks to express confused notions. It brings together phenomena the most diverse, and discovers the hidden analogies which unite them. If matter escapes us, as that of air and light, by its extreme tenuity, if bodies are placed far from us in the immensity of space, if man wishes to know the aspect of the heavens at successive epochs separated by a great number of centuries, if the actions of gravity and of heat are exerted in the interior of the earth at depths which will be always inaccessible, mathematical analysis can yet lay hold of the laws of these phenomena'. It makes them present and measurable, and seems to be a faculty of the human mind destined to supplement the shortness of life and the imperfection of the senses; and what is still more remarkable, it follows the same course in the study of all phenomena; it interprets them by the same language, as if to attest the unity and simplicity

> of the plan of the universe, and to make still more evident that unchangeable order which presides over all natural causes.
> (Fourier 2009, pp. 7–8).
> THE effects of heat are subject to constant laws which cannot be discovered without the aid of mathematical analysis. The object of the theory which we are about to explain is to demonstrate these laws; it reduces all physical researches on the propagation of heat, to problems of the integral calculus whose elements are given by experiment (Fourier 2009, p. 14).

Fourier's words are clear: it is possible to read the essence of nature, to distinguish essence and accident and to delimit every time a conceptual space where physical object is known definitively. In this scientific context, physical reality plays some fundamental gnoseological roles:

> Profound study of nature is the most fertile source of mathematical discoveries. Not only has this study, in offering a determinate object to investigation, the advantage of excluding vague questions and calculations without issue; it is besides a sure method of forming analysis itself, and of discovering the elements which it concerns us to know, and which natural science ought always to preserve: these are the fundamental elements which are reproduced in all natural effects (Fourier 2009, p. 7).

Therefore, physical reality plays the roles of *source* and *verification* in understanding results. Actually, there is another fundamental element emerging from these quotes: the properties of physics (that remains a "field" of calculus) play a role in mathematical scientists's work *legitimately*. Actually, they allow that "vague issues and the dead-end calculus should be excluded" because "pure" mathematics could fail. If differential equations are the language of nature and if physics is reduced to a field of calculus, it is just necessary to *formalize* physical problems in order to obtain some results. The *exactitude of mathematical procedures* and *physical certainty* in the existence of mathematical results can reach the *truth*. As a matter of fact, before and after Cauchy, scientists avoided to question on existence and uniqueness of solution in differential equations because they were basically presupposed. Hence, there is a particular relation between linearity and non-linearity.[12]

Before Poincaré, *linearization* was considered as *the* way to know the evolution of a trajectory *globally*, but not as an instrument allowing to work with non-linear and non-integrable equations. However, *linearity* becomes a fundamental concept of physics.

> Each observation has for an analytic expression a function of the elements which we wish to determine; and if these elements are nearly known, this function becomes a linear function of their corrections. In equating it to the observation itself there is formed an equation of condition. If we have a great number of similar equations we combine them in such a manner as to obtain as many final equations as there are elements whose corrections we determine then by resolving these equations (Laplace 1840, p. 74).[13]

[12] It should be clarified that speaking about non-linearity, the reference is to non-linearity and non-integrable equations.

[13] The first one italic is mine, the second one is by Laplace.

> Generally the errors of the results deduced from the great number of observations are the linear functions of the partial errors of each observation (Laplace 1840, pp. 305–306).

In this framework, perturbations (or noise) is considered as something which never determines a trajectory essentially.

> The phenomena of nature are most often enveloped by so many strange circumstances, and so great a number of disturbing causes mix their influence, that it is very difficult to recognize them. We may arrive at them only by multiplying the observations or the experiences, so that the strange effects finally destroy reciprocally each other (Laplace 1840, p. 73).

The basic idea is that a mathematical approximation can never cause the sensitivity to the initial conditions. Once again, this is the problem of measurement.

The reason of this is precisely explained by the particular philosophical relation between mathematics and nature.

The metaphor of program and its "ontologization" have already been mentioned; as regards physics, it is necessary to underline the coincidence between mathematics and nature and to seek its *philosophical* reasons. In fact, although there *were* non-linear differential equations and some physics problems producing chaotic dynamics since Newton and Leibniz times, only after almost two hundred years, Poincaré has started to talk about sensitivity to initial conditions in his work on the three-body problem (Marinucci 2011; Barrow-Green 1997).

It is necessary to look for philosophical reasons for this lateness in the physicists work. Previous scientists quotations show that they assumed that mathematics is nature and not only an instrument to interpret it.

The most philosophical argumentations are obviously those from Laplace. In his work the distinction between accidental and essential causes, involving the distinction between essential space and accidental space, can be found.

The parallel with biology is henceforth clear: the idea that the access to the measurement is exact implies that the effects of noise (or perturbations) delete themselves reciprocally. Hence, in this theoretical framework chance and determinism are mutually in contradiction: in other terms, determinism and predictability are reciprocally involved. Chance and chaos can have no theoretical space. Thus, there is always a linear proportionality between cause and effect, but this is just a *particular way* to interpret physical interactions. The concept of interaction or relation will be fundamental for leaving a deterministic perspective and opening up to a "relational" approach.

Interactions in Deterministic and Reductionist Framework

In the physical and biological frameworks outlined until now, a specific way to conceive interactions among the elements of a system emerges explicitly. In order to clarify this issue, it is important to stress that interactions have no real epistemological status and to show the limits of this approach.

In a reductionist framework, in order to know something, it is necessary to *isolate and define* some supposed simple elements, determining a closed space, in which possibilities are *totally* pre-determined. In other words, the sphere of possibility is *totally* determinable by the essential elements of reality. If it is necessary to isolate and to define, then interactions are only a simple *combinatory* among elements. Anyway, as a matter of principle, what interactions can produce is reduced to its essential elements and it is known *a priori*. This kind of knowledge is developed *from what is simple to what is complicated*[14] and it never arrive to get what is *complex*, neither emerging properties, even if *after* Poincaré the same elements can produce *emerging* properties. Thus, interactions have not any epistemological status. Knowledge is purely a combinatory, probability can be used in filling the space of possibilities. If it is just about combinatory, it can be affirmed that there is always a correlation between calculation and measurement. The effect of an interaction will be always a subset of the set of possibilities. The result of a throw of dice could be only one of six faces, one of the possibilities known *a priori*. In this sense, combinatory is enough to fill the list of possibilities, known *a priori*.

From the philosophical point of view, it can be said that orthodox biology and physical approach before Poincaré lead to the fundamental distinction between essential space and accidental space. In this framework, biologists and physicists arrive to presuppose that the access to measurement is exact, following different paths. On the biology side, the metaphor of program requires discrete mathematics; on the physics side, the reduction of non-linearity to linearity allows to know trajectories *as if* the access to measurement were exact. In these terms, noise does not play any epistemological role, it is in fact confined in the accidental space. The separation of these two kinds of space is theoretically fundamental, because it imposes that interactions of elements in a physical or biological system have to be *linear*. In other words, they will be guided by a linear causality which prevent that an accidental cause had an essential effect in any process.

Anyway, as a matter of principle, by knowing every essential cause of a process, it is possible to find out the Truth. When it is not possible, it is a matter of *epistemic* reasons. So, it is enough to use probability in order to measure the distance of Truth.

The Concept of Space as Container

The last concept, that has to be introduced, is the "space", understood as "container". This idea is implicit in what was outlined, but it is also useful to stress this topic.

The distinction between essential and accidental space has allowed to understand that possible phenotypes are given *a priori* in biology, starting from the combination

[14] The work of Poincaré about three body problem shows that it is not always possible to apply the same method. The three body problem is qualitatively different from the two-body. In this sense, it is not about a "complication" of two-body problem.

of their constituent elements. Therefore, the space of possibilities is given *a priori*. In fact, in a deterministic theoretical framework, it is possible to pre-determine every possible trajectory. Generally speaking, if the space of essential causes is known, its possible to develop its space of possibilities only combinatorially. In the orthodox biology, the elements of accidental space can play a role only in the choice of phenotypes, but never in their constitution.

In order to overcome this method of knowledge, it is necessary to debate on this concept of space from the physical, biological and philosophical point of view.

In the history of thought and science, there are two fundamental ideas of space. It can be conceive either as a "container" or starting from objects. The first representation was developed in particular by Newton and Kant, while the second by Leibniz, Mach and Einstein (Barbour 1989; Jammer 2008).

In orthodox biology, it is clear that the space of possibilities is not derived from its objects; on the contrary it is considered as a container. Specifying that the space of possibilities is more abstract than the one from Newton, it inherits anyhow some characters. So, space is motionless and absolute, likewise it is given *a priori* and it can not change.

Starting from the distinction between essential and accidental space and from the fact that interactions have not any epistemological status, it is possible to infer that the theoretical space of possibilities includes *all* possibilities.

The metaphor of container is today very present in physics.[15] It has allowed several scientific discoveries, but - like every metaphor - it prevents from conceiving certain kinds of properties and phenomena. In this sense, it is by no means clear that one could apply it everywhere without considering its viable applicability.

In the following chapter, the weaknesses of orthodox biology approach will be highlighted from the philosophical point of view and a new theoretical path will be proposed in order to elaborate a relational biology.

As reductionist and deterministic, orthodox biology imposes to conceive the space of possibilities *statically*, even if biological possibilities can change. The traditional biology is not able to conceive the intrinsic dynamics of life, because the only kind of variation that *can* be explained in its theoretical framework is reduced to DNA. The radical variation is outside of the theoretical framework and, therefore, it is not explained. The space of possibilities *does not change* starting from DNA's changes: it products a *new* space of possibilities. Before and after a change of DNA, there is no a changing space, but two different spaces, two different containers.

In conclusion, the following quote by Fourier summarizes the approach to nature that has been outlined here:

> The forms of bodies are infinitely varied ; the distribution of the heat which penetrates them seems to be arbitrary and confused; but all the inequalities are rapidly cancelled and disappear as time passes on. The progress of the phenomenon becomes more regular and simpler, remains finally subject to a definite law which is the same in all cases, and which bears no sensible impress of the initial arrangement (Fourier 2009, p. 8).

[15] The following chapter shows that the idea of relational space has been resumed from general relativity and that it is possible to apply it to biology, *mutatis mutandis*.

Bibliography

Bailly, Francis, and Giuseppe Longo. 2006. *Mathématiques et sciences de la nature*. Paris: Hermann.

Barbour, Julian. 1989. *Absolute or Relative Motion?* Cambridge: Cambridge University Press.

Barrow-Green, June. 1997. *Poincaré and the Three Body Problem*. Providence: American Mathematical Society.

Blay, Michel. 1992. *La naissance de la mécanique analytique*. Paris: Presses Universiter de France.

Boltzmann, Ludwig. 1905a. Über die Unentbehrlichkeit der Atomistik in der Naturwissenchaft. In *Populäre Schriften*. Leipzig: J. A. Barth.

Boltzmann, Ludwig. 1905b. Über statistischeMechanik. In *Populäre Schriften*. Leipzig: J. A. Bart.

Buiatti, Marcello, and Giuseppe Longo. 2013. Randomness and multi-level interactions in biology. In *Theory in Biosciences*, vol. 132, pp. 139–158. https://doi.org/10.1007/s12064-013-0179-2.

Deutsch, Jean. 2012. *Le gène. Un concept en évolution*. Paris: Seuil.

Fourier, Jean B. J. 2009. *The Analytical Theory of Heat*. Cambridge: Cambridge University Press.

Galilei, Galileo. 2005. Il Saggiatore. In *Opere*, vol. 1. Torino: UTET.

Israel, Giorgio. 2003. *La visione matematica della realtà*. Roma-Bari: Laterza.

Jacob, François. 1970. *La logique du vivant*. Paris: Gallimard.

Jammer, Max. 2008. *Concepts d'espace*. Paris: Vrin.

Kupiec, Jean-Jacques, and Pierre Sonigo. 2000. *Ni Dieu ni gène*. Paris: Seuil.

Lagrange, Joseph-Louis. 1788. *Méchanique analitique*. Paris: La Veuve Desaint.

Laplace, Pierre Simon. 1840. *Essai philosophique sur les probabilité*. Paris: Bachelier.

Longo, Giuseppe, and Maël Montévil. 2014. *Perspectives on Organism: Biological Time, Symmetries and Singularities*. Berlin: Springer.

Marinucci, Angelo. 2011. *Tra ordine e caos. Metodi e linguaggi tra fisica, matematica e filosofia*. Roma: Aracne.

Monod, Jacques. 1989. *Le hasard et la nécessité*. Paris: France loisir.

Wittgenstein, Ludwig. 1986. *Philosophical Investigations*. Oxford: Blackwell.

Relational Biology

The *theoretical* need of a relational biology is based on traditional biology's bias and limits. In particular, it is necessary to renew biology, using some ideas from contemporary physics and philosophy. It is very important to introduce a truly biological theoretical space, neither in contradiction with physics, but nor even reducible to the physical one.

It was debated, for example, about the impossibility of inscribing radical variation within the traditional theoretical frame, and about the epistemic status of noise, although non-linear differential equations are used in biology. In spite of this, variation should be considered an important element or, perhaps, the basic element of life. Variation is not only essential for evolution, it plays a fundamental role in increasing the robustness of the biological systems (Lesne 2008). Within a biological frame, even small variations are crucial at any level. In opposition to the classic "physics" biology (based on the fundamental biological Monod's invariant, DNA, and on a fundamental symmetry), relational biology considers radical variation as a theoretical principle. In other words, relational biology is not even based on a symmetry, but on *symmetry breakings* (Bailly and Longo 2006). Therefore, using Longo's works about symmetries in biology, the aim to a relational approach is to consider variation as basic "invariant".

From the epistemological point of view, it has been observed that *non linearity* does not always allow to discriminate a space of essential causes from the space of accidental causes, because of the sensitivity to initial conditions. It is about the fact that perturbations become crucial. In this sense, it is a matter of a kind of "*enchevêtrement des causes*", it needs to conceive biological elements differently as *distinct and inseparable*. Thus, it is necessary to go beyond the reductionist method with the aim to increase the possibilities of knowledge without methodological shifts.

In this sense, *relations* among elements of a biological system can have an own epistemological status; this means that some possibilities are generated by the interaction. In other terms, they are *not completely* predictable 'a priori', as it is

A. Marinucci, *Theoretical Principles of Relational Biology*,
Human Perspectives in Health Sciences and Technology 6,
https://doi.org/10.1007/978-3-031-39374-7_2

not only matter of combinatory. In fact, according to reductionist method, all the elements of a system should be always considered as *distinct and separable*. One could, perhaps, objecting that the impossibility of separating causes does not impede to interpreter the constitution of possibilities as a combinatory. The objection is justifiable, but it is based on a coincidence between the *concept* of randomness and the *mathematical tool* of probability: the reason why this conceptual coincidence should be abandoned will be shown below.

The problem of space is linked to these latter issues. Within the traditional biology, space is conceived as a container. In other terms, it is static, absolute and given 'a priori'. In this kind of conceptual framework, randomness can be reduced to probability and it becomes possible to hold the idea that biological possibilities (phenotypes) are constituted only combinatorially. In this sense, the space of possibilities is filled *completely a priori*. Nevertheless, biological spaces never stay the same, from tissues to the environment, even when complexity and elementary functions would not change. It is clear that it is necessary to conceive biological space differently from the physical one, adopted by orthodox biology. The essential problem is that living beings and environment act modifying mutually. Thus, it is necessary to introduce a concept of space that takes into consideration the *intrinsic* and typical transformation of biological spaces.

Accepting the idea that space actually changes and that it is, consequently, necessary to introduce the radical variation in the theoretical framework, it would be moreover difficult to conceive that the list of possibilities would be *completely* given 'a priori' and that the constitution of possibilities would simply be a result of a combinatory. The reasons of this statement will be explained, but for the time it is enough to remind that it would have been impossible to predict the existence of the ear bones simply starting by the jawbone of some Devonian vertebrate (Buiatti and Longo 2013). More in general, it is impossible to separate evolution from ontogenesis and it is necessary to keep always open the possibility of hopefulmonsters, as this is actually always possible, although undetermined, even when its 'probability' is next to zero.

The theoretical situation becomes even more complex, considering that in biology it is necessary to consider quantum processes (Del Giudice et al. 1986; Del Giudice and Preparata 1998; Del Giudice and Vitiello 2006). Quantum Mechanics does not allow to conceive causality like in classic physics: this involves the introduction of *contingency* within the biological theoretical frame.[1] This is not the only reason that forces to conceive a new role of causality[2] in biology. Some recent works (Buiatti and Longo 2013) have shown that there are several meanings

---

[1] In *Chance and necessity*, Monod states that microscopic processes are explained by quantum mechanics, however he never specifies that it is an intrinsically probabilistic physics and that it is in contradiction with classical physics.

[2] The fact that there is no necessity to consider relations causally will be explained. In this sense, the concept of "enablement" (Longo and Montévil 2013; Longo et al. 2012) will be used.

for 'randomness' in biology and these often overlay (each other). On the contrary of physics, in biology there is no coincidence between 'randomness' and 'probability'.

In order to solve these problems, it is impossible to remain within the theoretical frame of the orthodox biology, or better, it is necessary to abandon its hidden paradigm: the determinism of XVIIIth and XIXth. Generally speaking, contemporary biology does not rely too much on physics and philosophy progresses, even if it uses contemporary mathematical instruments and philosophical and physical concepts.

Beyond the necessity of renovating confrontation among biology, physics and philosophy, the science of life has to conquer a new and its own theoretical space.

Bailly and Longo (2006, pp. 243–246) have already shown that there is a crucial difference at a theoretical level between biology and physics about the interpretation of their objects and their trajectories.

If trajectories are specific and objects are generic within classical physics, on the contrary, trajectories are generic and objects are specific in biology. Considering physical objects, a falling object is always a generic falling object. In other terms, it is not necessary to specify the particular object from time to time. On the contrary, biological objects are specific, in the sense that it is not serially produced. Taking the mitosis as an example, reproduction creates always different cells. Moreover, this continuous differentiation (singularity of biologic object) is one of the most important invariant on which it is necessary to build a new biology. Once again, contingency becomes a new fundamental issue for biology.

Regarding trajectories, if in physics a geodesic is a necessary path for a deterministic theory and in relation to a fundamental symmetry, the opposite happens in biology, since each path is *contingent* and possible and characterized by symmetry breakings. The mitosis example has already been considered, but now it is necessary to underline the epistemological distinction between the domain of necessity, typical in classical physics, and the domain of possibility, typical in biology. Towards the end of this essay, this statement will be clarified from the philosophical point of view, but now it is necessary to briefly introduce this subject. Within classical physics, it is possible to determine each single trajectory necessarily, even in the non-linear case, in which two close trajectories would exponentially diverge. In other terms, surely there is unpredictability, even if *epistemic* as it is part of a deterministic frame. Hence, possibility remains conceived starting from reality. In biology, the general situation is different because it is impossible to conceive its possibility based on a physics model. Within biology, phenotypes are always different and, even if in some cases it is possible to foresee some phenotypes, it is not possible to know a *complete* list of them, as hopefulmonster is always possible.

In this sense, *each* biological object has its own evolutionary history, ontogenetically contingent, thus it is absolutely necessary to keep it safe introducing new theoretical elements.

In the next chapter, the opposition between the physical pair '*necessity-reality*' and the biological pair '*contingency-possibility*' will be explained.

It is necessary to introduce some theoretical concepts with the aim to supply the general (physical) guidelines towards a relational biology. In this sense, let's try to answer to the theoretical questions asked so far. The next chapter will be dedicated to the construction of the philosophical frame of relational biology.

The Concept of Relation and the Problem of Space

The physical metaphor of 'container' is not properly suitable and able to conceive an intrinsically changeable space as the biological one would be. Hence, it is necessary to introduce a different concept of space.

The idea of *relational space* can be found in philosophical and scientific traditions; it is namely possible to conceive it starting from objects.[3] In particular, this concept has been developed by Leibniz, Mach and Einstein. It would be extremely interesting to investigate more deeply this idea of 'space', *mutatis mutandis*, as this could easily apply to biology. This idea helps to avoid all the problems that the space as container implies.

Within this context, the relational space will be examined as it appears in general relativity. It is important to introduce a new idea of 'relation' with the aim to inscribe the biological interactions within a specific frame beyond the reductionism.[4] Obviously, it is important to show *how* the physical concept of relational space can be used in biology.

The Relation

Within the frame of classical biology and physics (before Poincaré) interactions have no epistemological status because it would always be possible to reduce their products to their constitutive elements.

Considering the limitation of this kind of interactions, it is necessary modifying the notion of relation. In general, the element of system will not be *necessarily* conceived as *distinct and separable* anymore. In this sense, the relation becomes crucial for and within biological processes.

If it is not possible to know 'a priori' the list of possibilities completely and if there is nothing outside of biological elements that can characterize a process, certain typologies of phenotypes will be enabled starting from *how elements actually interact* in a contingent context.

---

[3] It is very interesting to quote an excerpt by Leibniz about the concept of space. He wrote to Clarke on February 25th 1716: "For my part, I have said several times that I hold space to be something merely relative [. . .] taking space to be an order of coexistence" (Leibniz and Clarke 2007). See De Risi (2007) and Mugnai (1976).

[4] It is important to remind that it is not about removing reduction, rather radically limiting reductionism.

It is clear that relations produce some 'innovation' (i.e. phenotypes not predictable 'a priori', a hopefulmonster), that can be called 'specificity'. The combinatory is not able to give account of biological unpredictability.

Relation is not properly a cause, this represents the way in which, for example, cells are able to shape tissues. It is the way of "being together" and to let the elements of a biological system interact.

Hence, relation starts to acquire a theoretical status and a proper epistemological value. This is characterized by crossing of reductionism because this new relation will allow to describe unforeseeable phenotypes and, it is at the same time characterized by a breaking of the link between calculus and measurement.

In general, this kind of relation modifies biological *space* of possibilities and biological *possibilities*. So variation can be introduced in the theoretical frame. In fact, theoretical focus should be shifted from constitutive elements to relations, in order to be able to abandon the orthodox biology approach, going towards a relational biology. It is impossible to conceive relation only as combinatory.

If reductionism conceives the elements of a system as *distinct and separable*, within the relational frame, they are *firstly* considered as *distinct and inseparable*. The inseparability will open the doors of complexity, of specificity, of something that is *intrinsically beyond determination*.

If the elements of a system are distinct and inseparable, it is possible to open a theoretical space to conceive biological interactions as *never identical* iterate processes (Longo et al. 2012; Longo and Montévil 2013); theory is now able to conceive singularity of life. This has deep consequences on the concept of space as a container. Consequently, physics offers some possibility to conceive a relational space.

The Relational Space and General Relativity

The general covariance is the fundamental symmetry of general relativity. This actually expresses a kind of invariance in relation to diffeomorphic transformation.[5] Considering the covariance as the core of interpretation of general relativity, it is possible to reach the relational core of Einstein's ideas. The primary references are the works of Rovelli and Lachièze-Rey. They have drawn attention on the importance of general covariance with a truly relational interpretation of general relativity, referring directly to the hole argument (Rovelli 2010, pp. 68–71 and Macchia 2006).

[5] "A diffeomorphism is a deformation of space-time that moves all points arbitrarily. [. . .]. The original space-time is thus transformed into another space-time, different from the first, ma which is 'diffeomorphic' to it. The covariance of general relativity is expressed as the indifference of its laws to such deformations" (Lachièze-Rey 2008, p. 125). See also Rovelli (2010, p. 62).

During the elaboration of his theory, Einstein had to face the choice between two options: "(i) the field equations must not be generally covariant, or (ii) there is no meaning in talking about the physical space-time point A" (Rovelli 2010, p. 68). It is well known that Einstein chose the second option.

From this point of view, relations are truly the most important elements that characterize relativity. Concisely, the main elements in relativity are the physical fields and, in particular, the curvature tensors. In this sense,

> When I give the coordinates (position and date) of an event, I think I'm locating that event in space-time, but in fact I'm only locating the values of all the physical fields attached to that point (Lachièze-Rey 2008, p. 125).

In order to really understand the word "relativity", it is necessary to focus on elements and fields that live *on* or *in relation to* fields and not on elements (particles or fields) *in* the space-time.

> The theory does not predict what happens at spacetime points (like Newtonian and special-relativistic theories do). Rather, it predicts what happens at locations determined by the dynamical elements of the theory themselves (Rovelli 2010, p. 70).

This is the true aim and the effort to overcome the concept of space as a container given a priori. Relativity becomes "general" when it goes beyond the Newtonian space. This old concept is no more enough to describe an event because it is necessary to specify its physical context, in other words its relations. It is the theory that determines locations and, moreover, if space is not a mere container, talking about "empty space" becomes nonsense.

In this sense, considering a particle on a geodesic, it is possible to shift the reference of calculus and measurements of movement from the object to the gravitational fields and to their relations. In other words, each point of a geodesic can be conceived starting from objects or from relations. As regards biological fields, relativistic space represent finally the crucial point to conceive space, considering relations among objects. *Mutatis mutandis*, in biology it is possible to apply measurements and calculus, that usually concern the objects, to the characteristics of space. In this way, it could be possible to characterize biological spaces *directly* and biological objects *indirectly*. Moreover, the lack of a fundamental symmetry will imply some crucial consequences that will move away the relational biological space from the physical one. Before getting deeper on this issue, it is necessary to introduce the idea of relational space.

From Physical *Relational Space to* Biological *Relational Space*

The most interesting differences that have to be examined in depth about relational space of general relativity and biological relational space are: (1) the possibility of conceiving measurements and calculations on objects as determinations of their fields and relations; (2) how space is conceived starting from relations among objects. In fact, if space is conceived like Newton, calculations and measurements exclusively belong to the objects within[6] the space. Hence, it is a case of an absolute universal system of reference.

If a Newtonian space is not presupposed, it is necessary to conceive space starting from objects, *able to interact among themselves*. It is exactly the case of general relativity and relational biology.[7] This last way of conceiving space creates the possibility of introducing new conceptual possibilities to conceive a new biology.

In biology, relational space and the epistemological status of relations become extremely interesting for interpretation of laboratories results, that are very difficult to explain within an orthodox biological framework.

> *The tissue organization field theory of carcinogenesis* and neoplasia states that carcinogens disrupt the flow of the information between the stroma and the parenchyma and/or among cells within those tissue. The temporary or permanent effects of carcinogens on the intracellular structures and components, while variably deleterious to each of them, are not **directly** responsible for the development of a neoplasia.[...]. Hence, carcinogenesis and neoplasia may occur entirely through emergent (supracellular) phenomena once the signals that maintain normal organization are disrupted (Sonnenschein and Soto 2007, p. 117).

This kind of relational space allows to abandon a reductionist and deterministic theoretical frame. Actually, it gives a new perspective, focusing on the organization of biological systems at any level.

From the practical point of view, it is very difficult to trace the origins of biological space, but the inquiry on origin[8] is also not interesting, because it is a reductionist question. On the contrary, it is necessary to stress relations a bit more deeply. The idea that the elements and its environments can mutually modify themselves can certainly be applied to biology, but it is necessary to remind that biological objects are specific and that ontogenetic trajectories are possible paths, never necessary. I will be discussing the distinction between the biological possibility and the physical necessity, but now it is necessary to focus on biological space.

A biological space *is not constituted* by a relation among its elements (cells, for example). It *is* the relation among biological elements. It must be understood that space is never considered in itself because it is necessary to avoid falling into the "container" metaphor.

In a relational space, it is not possible to talk about an isolated biological object "into" the space, it is always necessary specify the context, i.e. the relations characterizing that specific object. For example, as regards a living being, "being into an environment" means "being in relation with." On the contrary, dying means leaving a relational biological space and entering in a physical space, hence, a

[6] "A place is the part of space occupied by the body and, depending on the space, can be absolute or relative" (Newton 2008).

[7] The fact that the interaction of biological elements is radically different from the one of physical objects will be shown.

[8] This concept will be analyzed in the last chapter of this book, see the section *The myth of "origin"*.

living being becomes a closed physical dissipative system. In biology, space can be conceived as an *order of coexistence*[9] of specific biological elements. Evidently, it is an order that can always be modified in relation to the changes of its elements and/or its relations. It is necessary to highlight that this is properly a characteristic of relational space. If comparing it with the idea of space as a "container", relational space is intrinsically changeable and dynamic, hence it has no substance in itself. Indeed, the process of mitosis shows that, even if a tissue is able to renew itself, its functions would *necessarily* need to change. Interactions, shaping biological spaces, are different from relativity. In fact, the parallel between biological relational space and physical space is very limited.

Gravitational interactions are represented in a deterministic theory, producing geodesics (necessary pathways). It is a set of interactions determined *causally*. In this sense, for example, the birth of a star modifies the causal relations of gravitational fields, but it is always inscribed in a *deterministic frame*, whose trajectories are the *new necessary pathways*.

The presence or absence of an object in the relational physical space is considered differently in biology.

It is precisely all these problems and theoretical elements, that put the concept of space towards a specific biological direction, that is irreducible to physics. The epistemological status of biological relations opens possibilities to conceive radical variation and *intrinsic* indeterminacy of some biological possibilities. In physics, on the contrary, relations are conceived and inscribed within a deterministic frame, admitting only an epistemic indeterminacy. Physical relations (for example in relativity) can be generally subsumed into a fundamental symmetry, the general covariance. On the opposite side, biology must take into consideration a fundamental indeterminacy that prevent from building a deterministic biology.

Biological relations and spaces are not directed by causal and linear grids. They are relations with their own epistemological status that "enable" to conceive radical variation. This latter is conceived then at the relations level, for example at the cell's organization level, and not at a lower level. This is not only a theoretical necessity (*de iure*), but also and especially an experimental necessity (*de facto*).

> A change in the appearance and/or behavior of a cell (phenotype) does not require a change in the structure of its DNA (genotype), but a change in the repertory of genes being expressed (epigenesist) (Sonnenschein and Soto 2007).

In fact, considering a biological relational space, it is about relations among biological elements (for example, cells in relation to tissues or plants in relations to environment). In this context, elements modify space and themselves as they interact. This is not the only way of modifying space. In general, biological elements reproduce themselves. Biological reproduction occurs always in a *specific* context, structured by relations enabling and canalizing reproduction.[10]

---

[9] Leibniz used this expression to define the concept of space conceived starting from objects.

[10] In the last chapter of this part, the crucial epistemological relevance of context and biological contingency will be underlined.

Considering Darwin's first principle[11] (the default status of a living being is "descent with modifications"), biological and physical constraints (Mossio and Moreno 2010; Mossio et al. 2013) play a fundamental role in the canalization of variation.

As for space, its relational structure is the crucial element that enables every time the production of specific elements, that can reproduce, or not, the very same relations. At this point, it is necessary to consider relational space as dynamic and never static, because this is always characterized by variations.

In this sense, for better characterizing the biological space and in order to show in which sense measurements and calculations *need* to be shifted from the biological object to the space, it is necessary to specify the distinction between randomness and probability.

This shows the necessity of focus the attention on the biological equilibria and organization, rather than on the element in itself.

Randomness, Probability and Measurement in Biology

Classic physics, not only relativity, is a deterministic theory. On the contrary, quantum mechanics is *intrinsically* probabilistic. As regarding biology, it is not a deterministic theory, as the physics before Poincaré. In fact, using non-linear differential equations, it is necessary to take into consideration all the scientific and philosophical consequences regarding sensitivity to the initial conditions. It is necessary to abandon completely the reductionist determinism (Laplace), even if it remains at the base of Monod's and Jacob's ideas. Nevertheless, this is not enough to conceive relational biology from the strictly probabilistic point of view[12] (Kupiec and Sonigo 2000; Kupiec 2012), it is necessary to acknowledge that in biology there are quantum (Buiatti and Longo 2013; Del Giudice et al. 1986; Del Giudice and Preparata 1998; Del Giudice and Vitiello 2006) and non-linear processes and, in addition, that biology presents its own characteristics, irreducible to physics. It has been shown the direction in which it is necessary to understand the concept of relation, i.e. from the consequences of a *properly biological* distinction between randomness and probability.

Contrary to physics, randomness and probability are not always coincident. In order to avoid misunderstandings, it is necessary to remind that "randomness" is a concept and "probability" is a mathematical instrument. In the biological context,

[11] Indeed, the second principle ("selection") presupposes a fundamental variation based on which selection is possible (Darwin 2011).

[12] Actually, Kupiec specifies that he does not make "in no way reference to a randomness analogous to the randomness of quantum theory, which would be constitutive of matter. The randomness I'm talking about remains linked to thermal agitation" (Kupiec 2012, p. 51). From the perspective of relational biology, this type of chance is only combinatorial and, therefore, it implies theoretically that it is possible to know a priori all the possibilities. This indeterminacy can only be *epistemic*.

it is important to remind that, in order to apply any physical or philosophical instrument or concept, it is always necessary to verify whether this is effectively possible. For example, probabilities can not be expressed if all possibilities are not given a priori. In biology, the space of *all* possible phenotypes is never given a priori. It is a crucial problem, as the sum of probabilities must be 1.[13]

In order to clearly understand the relation between randomness and probability in biology, it is possible to use *logically* a parallel, with the couple non-linearity/linearity. Before Poincaré the *linearization* was considered as one of the instruments that allowed to *globally* calculate the evolution of a trajectory. Summing up, there was an almost complete coincidence or superimposition between linearity and nature. After the "non-solution" of the three-body problem, the linearization has been considered nothing more than a possibility to study the non-linear differential equations. It has been understood that it is impossible to reduce non-linearity to linearity: the noise (often banished to run in background) has started to play a fundamental role. In fact, the change in the epistemological role of noise (or randomness) has led to the "deterministic chaos." After Poincaré, then, linearization has remained as an essential instrument for physics, but it is now *interpreted* differently. It will become clear later, that relation between randomness and probability is "similar" to the relation between non-linearity and linearity. This involves the necessity of a new interpretation of biological measurements and calculations.

In order to introduce the relation between randomness and probability, it is important to stress these concepts in physics. It has be shown that it is necessary to know a priori all the elements of space of possibilities and that the space of events has to be a subset of the space of possibilities[14] in order to apply probability to a random phenomenon. Let's consider a dice (or the spin up-down of an electron).[15] If it has six sides, the space of possibilities will not account more than six elements (numbers 1, 2, 3, 4, 5, 6), the probability of each one will be 1/6 and, finally, their sum must be 1.

In a physical context, it is possible to describe *totally* a random phenomenon through probability. As regards dice, only one probability can be applied to each side and the space of events could contain only one of possible elements and *nothing more*. If the space of possibilities "changes", it would be necessary to normalize, with the aim to attribute new values to probabilities. Considering *normalization*, the general situation does not change, as it is necessary to know a priori the new totality of possibilities and also the sum of their probability must be always 1. If these conditions are not verified, it is not possible to assign any value to each possibilities.

[13] This is a statement posing several mathematical problems, but which owns a strong biological meaning.

[14] The fact that space of events must be a subset of space of possibilities implies that, upon rolling a dice, it is impossible to get, for example, 11.

[15] It is actually possible to draw examples from classical physics or quantum mechanics because in both of them the complete list of possibilities is known a priori.

In this sense, considering normalization, it is not possible to properly talk about a space of possibilities, able to change. It is necessary to consider that there are two spaces of possibilities well *distinct and separated*, as the probabilities of elements in the two spaces (before and after normalization) are necessarily different. From the philosophical point of view, it is again a matter of the physical metaphor of space as a container. There are, indeed, two different containers, characterized by two totalities of different possibilities, whose probability values are never the same.

The biological situation is rather different: it is not possible to give a priori the *complete* list of possibilities and it is ever more problematic considering the space of biological possibilities as a not modifiable container. Hence, biology is characterized by an *intrinsically dynamic* space.

So, it would be at least difficult to apply probability to biological randomness. Now, it is about *specific biological randomness*, as its conditions are not the same as the ones of physics. Therefore, it is essential to take time to conceive the meaning of application of probability to biology.[16] Probability is and remains a valid instrument for biology, but not exactly based on the same model of physics. Indeed, *randomness is a concept, probability is*, on the contrary, *a mathematical instrument* allowing *an* access or *a* comprehension of randomness. This implies that probability can not provide a *complete* representation of biological randomness.

If space and changeable and unpredictable elements do not allow to use probability coherently, based upon the physics example, the proprieties of relational space offer different conceptual instruments to calculations and measurements in biology.

Talking about relational space, it is possible to consider calculations from objects and/or field perspective. This idea, *mutatis mutandis*, can be used to take into account biological random processes.

Without the *complete* list of possibilities (phenotypes), it is not possible to assign probabilities. Nevertheless, it is possible give a sense to measurements, when *environmental* conditions are *fixed* rigidly. In this case, it is possible to accept dangerous approximations, even though they are useful only for particular situations;[17] it is possible to talk about "probability" to predict some phenotypes. But "probability" related to what exactly? Related to the fact that hopefulmonster (i.e. what it is impossible to foresee) is *always possible*, that it can show itself in the *essentially incomplete* list of phenotypes.

It is clear that it is not possible to use such a concept of probability and that it is necessary to conceive differently biological measurements and calculations. Indeed, these will not directly produce probabilities and information on the objects, but on the environment's stability.

---

[16] Based on what it was be just said, Kupiec's idea of a probabilistic biology, like thermodynamics, becomes at least problematic (Kupiec 2012).

[17] In physics linearization can indeed be used, but only if (Lyapunov exponent) $\lambda \leq 0$ and considering Lyapunov's time.

The impossibility of predicting hopefulmonster, or predicting the moment in which this will show itself, and the necessity to always maintain open the possibility of its existence, make impossible to know a priori the *complete list of possibilities*. By this use, the notion of probability loses its accuracy.

Instead of probabilities and biological objects, let's consider and apply measurements and calculations on environment (space). If it is not possible to know the totality of possibilities, it is also impossible to talk directly about them. Here, relational space shows all its power and potentiality.

The new situation is characterized by a biological random phenomenon (for example, the constitution of possibilities), whose measurements *do not directly* describe possible phenotypes, that are within the *incomplete* list given a priori. Likewise they do not describe phenotypes impossible to predict (hopefulmonster). In brief, measurements *can not directly* talk about biological *objects*.[18] Measurements and calculations, normally applied to the possible biological objects, *can* be applied also to space *as it is relational*. In this sense, if the environment is well fixed, in relation to its history, it can be said that the *thickness*[19] of possibilities of predictable phenotypes is greater than the thickness of possibility of hopefulmonster.

Calculations referred on objects talk indeed about the "stability" of biological space, intrinsically changeable and able to produce known phenotypes and phenotypes never included in the a priori list. The impossibility to know the complete list of biological possibilities enables to apply probability of biological randomness. As regards application of measurements and calculations, relational space allows a shift from biological object directly to environment (biological space). These describe only indirectly its biological objects, but *directly* the stability of biological relational space (equilibrium of tensions).

In brief, it is possible to *directly* talk about relational structure of biological space of possibilities and only indirectly about its possibilities. In other terms, measurements and calculations talk about how a configuration of relational space enables the creation of possibilities within the *incomplete* list known a priori. In fact, relational space "enables"[20] and can not directly be cause of production of particular phenotypes, as there are always unpredictable phenotypes, equally "enabled" by the biological relations. In this sense, the stronger the relational space stability is, the

[18] For different reasons, quantum mechanics calculations prevent from referring directly to objects.

[19] A section of next chapter will be devoted to this concept, introduced here just intuitively. Thickness of possibilities will be defined as the measurement of influence of contingency over the sets of biological possibilities (predictable and non predictable) such that the latter ones become liable to enter into domain of reality.

[20] Regarding the difference between the notions of enablement and causality, see Longo and Montévil (2013): "In short, a niche enables the survival of an otherwise incompatible/impossible form of life, it does not cause it. More generally, niches enable what evolves, while evolving with it. At most, a cause may be found in the "difference" (a mutation, say) that induced the phenotypic variation at stake, as spelled out next".

greater is the thickness of possibility of predictable phenotypes as well. Thickness depends on the configurations of biological relations in their own contingency.[21]

Starting from the example of the relation between linearity and non-linearity, the role of probability can be interpreted in relation to the biological randomness. In this regard, *in biology* there is no coincidence between randomness and probability. However, there are cases in which it is actually possible to use probability. As in the example of the relation between linearity and non-linearity, now probability will only shed light on some particular aspects of biological randomness processes. It will allow studying them, without being possible to reduce randomness to probability. It has been observed that, from the theoretical point of view, probability can not be used in biology, but measurements can be applied to the space, only considering the space as relational. Hence, possibilities can be coherently characterized using the concept of thickness.

Space stability and observations, always a posteriori, allow canalization of possibilities towards the most observed phenotypes, even if it is necessary to keep open the possibility of hopefulmonster. In this sense, it is possible to keep variable the thickness of possibility.

From the experimental point of view, there is a rigidly fixed environment, characterized by a phenotypical well canalized *history and* the *possibility* of hopefulmonster (i.e. of the radical variation) is open. From the theoretical point of view, it is an environment characterized by a contingent history and the possibility of the hopefulmonster. The *historically situated* stability and the environment *measurability* give the thickness of possibility.

At this point, it is necessary to clarify a very important aspect. Generally speaking, the theoretically most important element is that, from the point of view of possibility, it is not directly a matter of objects. Indeed, it is necessary to clarify that hopefulmonster is not being *directly* appointed here, because it is undetermined and indeterminable from the qualitative and quantitative point of view. *Possibility* is what is being considered, namely relations that *can* imply a certain thickness.[22] It is almost the same for historically predictable possibilities. In fact, it is impossible to *quantitatively* predict the thickness of all them. Hence, it is only possible to *directly* talk about the *sets* of predictable and unpredictable possibilities.

If biological space is namely the relation among elements, these same elements refer themselves to relations stability, *historically situated*, to a particular configuration of organization, to the *equilibrium of tensions* characterizing biological levels of living beings systems.

It is evident that this kind of consideration is based on an a posteriori knowledge that can not deny the *possibility* of a radical variation. This way of presenting these

[21] Concerning the relation between possibility and reality in biology, see next chapter.

[22] It is very important to emphasize the notion of *possibility*. The next chapter shows that biology is immersed in the realm of the possibility and the concept of "thickness" allow to shift from the realm of possibility to the reality one.

themes is utterly necessary, if it takes into consideration and definitely safeguards the singularity of biological objects.

Physical Symmetries and Biological Breakings

The introduction of relational space allows to overcome the problems of the biological distinction between randomness and probability and, moreover, the problems concerning the use and the interpretation of measurements and calculations in biology.

Now, there is a theoretical aspect really important that needs to be discussed. It underlines a deep difference between the application of relational space to biology and to physics.

It is necessary to highlight that, if in physics it is possible to consider the calculations and measurements by the side of the object *and* by the space; in biology this is not possible.

In this sense, it is necessary to shed light on the theoretical reasons that prevent from directly talking about objects based upon the classic physics model. It has been seen, that it is possible to apply the concept of relational space to relativity on the basis of a fundamental symmetry, making acceptable the two interpretations here sketched. Saying that equations are covariant means that there is always something that is preserved. There is, hence, a sort of conceptual unity under which it is always possible reconduct all the transformations to. Also concerning biology, there should be a fundamental invariant. On the contrary, including the radical variation within the biological theoretical frame, it is necessary to abandon every deterministic and reductionist approach (like that of Monod and Jacob). Furthermore, a merely probabilistic perspective is not enough to find an invariant, because probability is not able to completely give account for the biological randomness. The particular relation between randomness and probability creates the necessity to talk directly and exclusively of relational space, of its stability[23] and of the *sets* of predictable and unpredictable possibilities.

Following Darwin's ideas, embraced by Longo, Soto and Sonnenschein, it is finally possible to consider *variation* as the fundamental invariant of biology. The situation is now the following. Considering variation as the fundamental invariant, biology is characterized by *symmetry breakings* (Bailly and Longo 2006; Longo and Montévil 2014), at the level of elements and space production. It is important to clarify some concepts. From the explicative point of view, it is necessary to avoid misunderstanding between breakings regarding the production of biological

[23] Moreover, it is possible to increase the stability of a relational space in order to increase the thickness of introducing certain phenotypes. In this sense, it is possible to talk about processes canalizing the stability of biological space. The fact is that, with respect to biology, this issue develops in the only direction that leads from elements to space... from cells to tissue etc.

elements (for example, cells) and breakings regarding biological relations. It is known that mitosis is characterized by symmetry breakings. In fact, cells are never produced in series. The fact that these are all made different does not necessarily imply that their relations are completely causal. Let's not forget that cells always arise and are *already* situated inside specific (*contingent*) relations, characterized by an own epistemological status, and that each and every biological element is what it is in virtue of its ability of interacting. Cells can, then, maintain or break the pre-existent relations, implying, in this latter case, a deep change of its relational space, which can become able to produce cells with different functions.

In this sense, starting from the experimental point of view, Ana Soto and Carlos Sonnenschein postulate "an alternative theory [they] call the *tissue organization field theory of carcinogenesis and neoplasia* which posits that neoplasias are emergent phenomena resulting from flawed interaction among cells and tissues" (Sonnenschein and Soto 2007, p. 97).

Relations are able to shape what could be called a "*equilibrium of tensions*", never necessary, and always needing to renovate itself. Through the process of renovation, this equilibrium can change and imply a symmetry breaking at a space or tissue level, for example producing cells, different from the functional point of view. Equilibrium changes with regard to cells complexity and function, *always because its ability of interaction*. In this sense, there is no necessary equilibrium in biology, because biological elements are specific and its relations do not fall under the domain of reductionism. At this point, the fundamental epistemological role of relations emerges strongly: there is no contradiction between the concept of symmetry breaking and the production of order.[24] For what concerns space description, the fact that relations have its own epistemological status implies that it is necessary let them "play" (allow them to be) in order to be able to understand, a posteriori, if the built space changes or remains stable, while undergoing the process of renovating itself. In this sense,

> *In culture*, cells from metazoa show properties that they do not demonstrate at the organismal, organ, or tissue levels. By virtue of becoming freed from the needs of a multicellular organism, the 'liberated' metazoan cell *in culture* may re-acquire ancestral, cryptic properties, including proliferation and mobility (Sonnenschein and Soto 2007, p. 80).

The game of biological relations takes place between variations and its constraints. If space is relational, then it is possible to conceive constraints starting from relations. In other terms, context is nothing more than the pre-existent contingent structure of relations that determine the contingency in all biological processes. Conceived starting from *distinct and inseparable* elements, relations are so determined that the couple variation-constraints is not necessarily perceived as an opposition. In fact, variation can definitely contribute to augment the robustness of a biological equilibrium of tensions (Lesne 2008).

[24] It is necessary to underline that this approach is very far from Monod's dichotomy between chance and necessity, according to which order can only come from order.

Even though these themes are worth to be better developed, there is at least a theoretical "place" to constraints[25] because, as variation plays a fundamental role, it is necessary to explain the possibility of stability, of the fact that in biology equilibria can be kept. In this sense, the stability of biological space implies necessarily a number of webs of constraints that can interact with variation, in order to keep themselves. In general, it is being talked about relations and equilibria that can *chronologically* happen, for example, cells produced by mitosis.[26] Each and every cell will apply to constraints (for example inside the tissues) its own specificity and will interact randomly, not at all deterministically, but it will fall within a well constraint and contingent context. On the basis of different biological elements, *as interacting themselves*, relational structures will stay the same or will change.

> When a single cell of these different cell types was transplanted into mice, only the teratocarcinoma stem cells gave rise to neoplasias. The differentiated cells derived from these stem cells did not form tumors. Thus, genuine neoplastic cells were able to generate normal cells and normal tissues. These findings contradict the concept that the neoplastic, malignant behavior is 'fixed' (Sonnenschein and Soto 2007, p. 125).

This is exactly the fundamental point for a relational biology, in which elements are considered, first of all, as *distinct and inseparable*.[27]

Certainly, at a relational level, each tissue (or space) has its own parameters and its own observables, in order to mathematically determine its stability in relation to its history. But there is always a relational game, never completely predictable, for stability preservation or symmetry breaking. On principle, each tissue *enables* different typologies of cells. These remains always *possible*, but from *reality*'s point of view, space or constraints stability makes more or less "*thick*" its own possibilities. Indeed, in biology there is no equilibrium preserving itself *necessarily*: it depends on relations, on interactions (Noble 2006; Bailly and Longo 2006).

Constraints push the equilibrium of tensions to maintain coherent its relations. In fact, each mitosis or each reproduction process occurs always in a contingent context, and its coherence can be broken or not.

It is evident that in biology there is no *necessary* symmetry and that life is characterized and renovated by breakings. In particular, the fact that it is not possible to predict a priori any symmetry breaking (Longo and Montévil 2013) and the fact that these are always possible (but never determinable) imposes the impossibility of talking *directly* about biological possibilities. Then, in order to understand the importance of symmetry breakings and their unpredictability, it is necessary to consider that calculations and measurements are referred to biological space. In this sense, it is necessary to talk about *enablement*, not about causality and not directly about possible elements of a biological system.

[25] In the next part, these concepts will be deeply developed.

[26] It is important to remind that, generally speaking, equilibria and constraints are structured by networks of changeable relations.

[27] One part of next chapter will be devoted to the dual role of biological contingency.

It has been shown the example of mitosis and tissue: cells are never identical and have a relational structure, but this issue can be extended to different biological levels. The fact that variation has been proposed as biological invariant leads to some considerations that divides definitely biological relational space from the physical one. At this point, the distance between the two kind of spaces becomes clear. The main point is the impossibility to apply biological measurements *directly* upon the objects.

In this sense, in biology it would be more than legitimate to talk about a relational space, but for different reasons from physics.

As shown, the theoretical crucial effort here is exactly being able to conceive variations, supposing that no unity is capable to understand and subsume variations completely and a priori. From the philosophical point of view, it can be considered the Wittgenstein's concept of language-games. He states:

> And this multiplicity is not something fixed, given once for all; but new types of language, new *language-game*, as we may say, come into existence, and others become obsolete and get forgotten. [. . .]. Here the term 'language-game' is meant to bring into prominence the fact that the speaking of language is part of an activity, or of a form of life (Wittgenstein 1986, §23).

In brief, as for language and biology, it is necessary to underline that activities of life and of language are always beyond representations or interpretations that can be given. From the philosophical point of view, it would be important to safeguard this aspect. Summarizing so far, the fundamental invariant of biology is variation, however, this is unpredictable. Even though, it is necessary to safeguard variation, as it is nonetheless the motor of life. With this aim in mind, it is necessary to concentrate on possibility and its thickness, on *enablement* and on relational space.

In this theoretical context, symmetry breakings are definitely a crucial element for relational biology. It has been seen that, starting from a fundamental symmetry (Monod), it is impossible to conceive radical variation within theoretical frame of biology, because this can not be led back within the limit of determination.

In general, the concept of symmetry breaking can be applied to singular biological elements (i.e. cells) and to space. The first case refers to ontogenesis, never as a necessary process, but always as a contingent process, characterized by constraints. The developmental process of a singular element (cell), referred to its relational space, is characterized by "cascades of symmetry breakings" (Longo and Montévil 2014). The second case refers to the fact that it is possible to talk about breakings not only in the case of individuals, but also in the case of space. This concept has already been simplified discussing equilibrium of tensions. Therefore, in biology is not necessary that an equilibrium stays stable meanwhile undergoing a process of renovation. A possibility of changing for complexity and function always exists. The introduction of the idea of relational space allows, indeed, to maintain indissolubly linked biological elements and space that becomes, ultimately, *intrinsically* dynamic and changeable.

Thus, biology is characterized by a status of "extended criticality" (Bailly and Longo 2006; Longo and Montévil 2013, 2014). Contingency and unpredictability are revealed to be fundamental elements to conceive biology as relational. Biological elements have an identity insofar as they are able to interact with themselves and with interactions (i.e. relational spaces) in which are immersed. In this sense, the role of constraints is crucial to canalize variation and to maintain biological structures, or better equilibrium of tensions. At this point, constraints find their theoretical space. In fact, biological equilibria and trajectories are never necessary. They can be broken or simply perturbed. It is always a matter of a contingent interactions.

Bibliography

Bailly, Francis, and Giuseppe Longo. 2006. *Mathématiques et Sciences de la Nature*. Paris: Hermann.

Buiatti, Marcello, and Giuseppe Longo. 2013. Randomness and multi-level interactions in biology. In *Theory in Biosciences*, vol. 132, pp. 139–158. https://doi.org/10.1007/s12064-013-0179-2.

Darwin, Charles. 2011. *L'origine delle Specie*. Torino: Bollati Boringhieri.

De Risi, Vincenzo. 2007. *Geometry and Monadology*. Basel/Boston/Berlin: Birkhäuser.

Del Giudice, Emilio, and Giuliano Preparata. 1998. A new QED picture of water: understanding a few fascinating phenomena. In *Macroscopic Quantum Coherence*, ed. E. Sassaroli et al. London: World Scientific.

Del Giudice, Emilio, and Giuseppe Vitiello. 2006. Role of the electromagnetic field in the formation of domains in the processof symmetries-breaking phase transitions. In *Physical Review A* 74 (2).

Del Giudice, Emilio, Silvia Doglia, et al. 1986. Electromagnetic fiel and spontaneous symmetry breaking in biological matter. *Nuclear Physics B* 275.

Kupiec, Jean-Jacques, and Pierre Sonigo. 2000. *Ni Dieu ni Gène*. Paris: Seuil.

Kupiec, Jean-Jaques. 2012. *L'ontophylogenèse*. Versailles: Quæ.

Lachièze-Rey, Marc. 2008. *Au-delà de l'espace et du Temps*. Paris: Le Pommier.

Leibniz, Gottfried Wilhelm, and Samuel Clarke. 2007. *Exchange of papers between Leibniz and Clarke*, ed. by Jonathan Bennet. http://www.earlymoderntexts.com/pdfs/leibniz1715_1.pdf.

Lesne, Annick. 2008. Robustness: confronting lessons from physics and biology. In *Biological Review* 83.

Longo, Giuseppe, and Maël Montévil. 2013. Extended criticality, phase spaces and enablement in biology. In *Chaos, Solitons and Fractals*. Amsterdam: Elsevier.

Longo, Giuseppe, and Maël Montévil. 2014. *Perspectives on Organism: Biological Time, Symmetries and Singularities*. Berlin: Springer.

Longo, Giuseppe, Maël Montévil, Stuart Kauffman 2012. No entailing laws, but enablement in the evolution of biosphere. In *GECCO Companion '12*. New York: AMC.

Macchia, Giovanni. 2006. *L'*Argomento del buco *di Einstein. Il recente dibattito sull'ontologia dello spaziotempo*. http://www.uniurb.it/Filosofia/isonomia/2006macchia.pdf.

Mossio, Matteo, and Alvaro Moreno. 2010. Organisational closure in biological organisms. In *History and Philosophy of the Life Sciences* 32.

Mossio, Matteo, Leonardo Bich, and Alvaro Moreno. 2013. Emergence, closure and inter-level causation in biological systems. In *Erkenntnis*, vol. 78, pp. 153–178. https://www.bing.com/ck/a?!&&p=6da8edfde86656bfJmltdHM9MTY5MzM1MzYwMCZpZ3VpZD0yNDE1MjQ0YS1mYmViLTY1NzItMzFkNi0zNWIwZmFmZjY0MmYmaW5zaWQ9NTU0OQ&ptn=3&hsh=3&fclid=2415244a-fbeb-6572-31d6-35b0faff642f&psq=Mossio%2c+Matteo%2c+Leonardo+Bich%2c+and+Alvaro+Moreno.+2013.+Emergence%2c+closure+and+inter-level+causation+in+biological+systems.+In+Erkenntnis.&u=a1aHR0cDovL2R4LmRvaS5vcmcvMTAuMTAwNy9zMTA2NzAtMDEzLTk1MDctNw&ntb=1"10.1007/s10670-013-9507-7.

Mugnai, Massimo. 1976. *Astrazione e Realtà: Saggio su Leibniz*. Milano: Feltrinelli.
Newton, Isaac. 2008. *Principi Matematici della Filosofia Naturale*. Milano: Arnoldo Mondadori.
Noble, Denis. 2006. *The Music of Life*. Oxford: Oxford University Press.
Rovelli, Carlo. 2010. *Quantum Gravity*. Cambridge: Cambridge Universit Press.
Sonnenschein, Carlos, and Ana Soto. 2007. *The Society of Cells*. New York: Taylor & Francis.
Wittgenstein, Ludwig. 1986. *Philosophical Investigations*. Oxford: Blackwell.

Preserving Possibility

So far attention has been drawn to the necessity to conceive newly the relation between biology and physics, in order to create an epistemological and theoretical area, properly referred to biology. It has been stressed problems that are difficult to solve within the theoretical frame of orthodox biology, principally starting from works of Bailly and Longo (2006), Longo and Montévil (2014), and Sonnenschein and Soto (2007). Let's propose possible paths on which it would be possible to find some solutions. In this sense, it has been shown that the main theoretical elements are the epistemological status of relation and the introduction of the idea of relational space. A coherently shift of measurements from biological objects to space has been allowed by these elements. Thus, it is possible to conceive the possibility of hopefulmonster, although indeterminable. More in general, a new approach has been proposed with the aim to *conceive* radical variation in a coherent theoretical frame.

At this point, it is the moment to clarify fundamental concepts, which represent the basic steps of everything that have been proposed, and that so far have not yet been examined in depth: the concept of *contingency*, *thickness* and the *transition between possibility and reality*. In this first part, these concepts will be examined from the relational space point of view; in the next part of this book, they will be analyzed from the perspective of biological times and organizations. By the analysis of these concepts, the theoretical difference between physics and biology will be clarified.

In this sense, it is important to explore the relation between two conceptual couples: *necessity-reality* and *contingency-possibility*. The aim is to show some essential elements for a new epistemological frame of biology. Specifically, a *philosophical* elaboration of a biology based upon symmetry breakings leads to a possibility to conceive "*differences without concept*", impossible in physics and in a *physical* theory of biology (Monod).

A. Marinucci, *Theoretical Principles of Relational Biology*,
Human Perspectives in Health Sciences and Technology 6,
https://doi.org/10.1007/978-3-031-39374-7_3

Physics

In the deterministic theoretical frame of classical physics, trajectories (as in general relativity) are geodesic, namely optimized and necessary paths starting from well-defined initial conditions. It is possible to talk about determinism because indeterminacy is *epistemic*, because it is produced by sensitivity to initial conditions, is epistemic. From the *mathematical* point of view, this means that this kind of indeterminacy depends on the problem of access to measurement.

From the physics point of view, in a non-linear system, it is certainly possible to separate different kinds of causes like Laplace, but it is always about relational structure in which causality plays a fundamental role, not just theoretical. In fact, given certain conditions, each single trajectory is fixed. What can not be fixed is the exponential divergence between two or multiple trajectories: there is always an error beneath measurement, which prevents predictability. Nonetheless, each trajectory remains ultimately *a possible path in the realm of reality*. In other terms classical physics describes *directly* nature through calculus. In this sense, it describes a possible *reality* or a set of *possible realities*. Indeed, considering all the possible trajectories of a phenomenon within their space of phases, it is not possible to talk about a set of possibilities, which can become real: it is a matter of possible *realities*. The crucial point is that possibility is conceived starting from reality and it becomes *flattened* upon reality. From the philosophical point of view, physics is definitely in the realm of reality. The fact that a deterministic physics can only contemplate necessary trajectories implies the impossibility to consider itself in the realm of possibility. If there is no breaking between calculus and measurement, all the trajectories are considered as necessary; they are anyhow *real*, even though they are not all necessarily *actual*. If trajectories, generated by a non-linear differential equation, are real, measurement allows to reach the *actual* trajectory, i.e. *hic et nunc*.

It is certainly possible to substitute the couple "possible reality-actual reality" by "possibility-reality", but the level of knowledge should be not confused with the level of the actual existence of the object.

It is a matter of a fundamental distinction already asserted by Kant (1996)[1] and Tagliagambe (1991) between how the objects are in relation with knowledge and the existential conditions of its existence. On the one hand, the question is *how* an object is in relation to knowledge,[2] on the other hand, the question could be: what is object? It is necessary to admit that in classical physics it is not easy to fully understand this distinction. Within a deterministic frame, physics refers *directly* to its objects, in the sense that physics usually establishes *what is* its object. Quantum

[1] Compared to what stated by Kant, this distinction is used quite differently, but it is important explicit sources (Tagliagambe 1991).

[2] In this context, it is possible to talk about object and knowledge, because the reference is just classical physics. Dealing with quantum mechanics and biology, it is more correct to speak about relations or interactions among systems. About a relational point of view applied to quantum physic, see Rovelli (1996).

mechanics helps to understand this issue. In this sense, Heisenberg underlines an essential distinction between two ways of conceiving the concept of realism.

> Practical realism assumes that there are statements that can be objectivated and that in fact the largest part of our experience in daily life consists of such statements. Dogmatic realism claims that there are no statements concerning the material world that can not be objectivated. [...]; actually the position of classical physics is that of dogmatic realism. [...]. Especially in [classical] physics the fact that we can explain nature by simple mathematical laws tells us that here we have met some genuine feature of reality, not something that we have—in any meaning of the world—invented ourselves. [...]. But quantum theory is in itself an example for the possibility of explaining nature by means of simple mathematical laws without this basis [dogmatic realism] [...]. Natural science is actually possible without the basis of dogmatic realism (Heisenberg 1971, pp. 82–83).

This distinction is similar to the one between reduction and reductionism, proposed in the first chapter. The practical realism is a scientific and philosophical tool that *can* be applied, for example, to classical physics, but this does not imply dogmatic realism, namely the idea for which the objectification of scientific assertions is necessary. It is not necessary that physics *should* talk *directly* about nature.

> Piratical realism has always been and will always be an essential part of natural science. Dogmatic realism, however, is, as we see it now, not a necessary condition for natural science (Heisenberg 1971, p. 82)).
>
> But the change in the concept of reality manifesting itself in quantum theory is not simply a continuation of the past; it seems to be a real break in the structure of modern science (Heisenberg 1971, p. 29).
>
> The mathematical symbols, describing such an observational situation, represent possibility rather than reality. Perhaps it could be said that they represent something intermediate between the possibility and the reality (Heisenberg 2010, p. 214).

These Heisenberg's quotations are truly important as they highlight the central differences between deterministic theoretical frame of classical physics and intrinsically probabilistic frame of quantum mechanics. This latter can only *indirectly* talk about nature, because there is a breaking between calculus and measurement (Bailly and Longo 2006). Quantum mechanics does not talk about trajectories *in relation to its object*, but rather about a wave-function of probability. From the philosophical perspective, quantum mechanics does not answer to the same question of deterministic physics. It describe *how* two or more systems *interact* to each other. What has been said is a crucial theoretical aspect. In order to conceive de-coherence, it is essential to consider common and microscopic coordinates as distinct and inseparable. It is not sufficient to *juxtapose* them, it is necessary to keep them together. In the first case (juxtaposition), it is possible to write as follows:

$$H = H_c + H_e$$

where H is a system and H_c and H_e represent respectively its macroscopic and microscopic part. It is easy to notice that a simple juxtaposition does not allow to *conceive* any kind of exchange of energy: H_c and H_e are firstly isolated and afterwards put together. On the contrary, if H_c and H_e are considered, first and foremost, as distinct and inseparable, it is possible to account of an effect of thermal dissipation. In this sense, it can be written

$$H = H_c + H_e + H_{int}$$

where H_{int} is dissipation. “Even if it happens that the environment wave functions for two macroscopically different states are coherent at some time [...], they will become very rapidly orthogonal because of their coupling with different values of the collective macroscopic observables”[3] (Omnés 1994, p. 269).

In other terms, not only possibility, but also the passage from possibility to potential or possible reality are built starting from interaction, admitting its own epistemological status. For example, through calculus it is possible to determine the possible status of the Schrödinger’s cat before the box is open. Among possible status, superimposition is a state that can not enter into reality.[4] In this sense, possibility starts to detach from reality. The case of classical physics is completely different, because all the possibilities reside in the realm of reality. In quantum mechanics, *it is not necessary that what is possible is intended to its realization.*[5]

Out of a deterministic perspective, it becomes possible to separate possibility from reality.

It is clear that, what has been affirmed, is not at all sufficient for completely characterizing these matters and, in particular, quantum mechanics. Nevertheless, it is enough to have underlined that a deterministic science and a probabilistic science answer different questions. Moreover, they can supply two different perspectives, even contradictory, on nature and on the relation between possibility and reality. In this sense, unlike classical physics, contingency, a-causality and indeterminacy become constitutive aspects of quantum mechanics. Accepting that Bell and Aspect have given a mortal strike to the hypothesis of hidden variables, it is not possible anymore to propose *necessary* paths in quantum mechanics. Hence, the latter poses a problem, not only in philosophical terms: the couple *contingency-possibility* has to be conceived radically. The need of conceiving this couple becomes stronger and stronger in biology. Before talking about biology, a scheme is useful to resume the differences between classical and quantum physics on possibility and reality.

---

[3] It is quite evident that in this context the word “coupuling” highlights the crucial importance of relations.

[4] the “multiple worlds” interpretation will not embraced, as it tries to keep a strong concept of reality, but without providing elements to retrieve other realities.

[5] The indeterminacy of totality of biological possibility forces to conceive the experience of a failure, which implies to investigate life in depth.

Classical Physics

$$\text{Realm of reality} \begin{cases} \text{Possible or potential reality} \\ \text{Actual reality } (\textit{hic et nunc}) \end{cases}$$

Quantum Physics

$$\text{Realm of possibility} \left\{ \text{Quantum possibility (quantum coherence)} \right.$$

$$\text{Realm of reality} \begin{cases} \text{Possible or potential reality} \\ \text{Actual reality } (\textit{hic et nunc}) \end{cases}$$

Biology

It has been taken into consideration a long theoretical path from the physics of Laplace to the biology of Monod and Jacob. Now, the target is a new biology, outlining a *relational* biology starting from the ideas presented in the book *Mathématiques et science de la nature* (Bailly and Longo 2006). Many underlined aspects can not be explained in full by traditional frame of biology or are not even taken into consideration.

At this point, the theoretical frame of biology can be essentially neither deterministic nor probabilistic (Monod 1989; Jacob 1970; Jean-Jacques Kupiec 2012). About actual reality, if classical physics (and deterministic biology) can admit only an epistemic indeterminacy and if quantum mechanics can admit only an epistemological indeterminacy (related to the *transition* from the realm of possibility to the reality one), biology admits an epistemological indeterminacy related to *possibility* and reality.

Compared to theoretical physical models, biology shows some aspects of irreducibility: the crucial role of variation, in all its shapes, the fact that trajectories are generic and the objects are specific and, eventually, the impossibility to know a priori the *complete* list of possibilities.

Considering these latter issues as essential for a real understanding of biology, it is necessary to emphasize the couple *contingency-possibility*, but differently compared to quantum mechanics. In physics, possibility is always determined a priori and, consequently, randomness can even coincide with probability. In biology, the role of contingency and the importance of possibility are even stronger. In fact, the typically biological relation between the concept of randomness and the mathematical tool of probability can not be forgotten nor the switch from space as a container to relational space, nor, finally, the switch from objects to space, considering biological measurements.

Biology is deeply set in the realm of possibility. In fact, biology has to keep into account a possibility that can not be foreseen. Indeed, in biology the connection between possibility and reality is never necessary, but always contingent. This means that, if biological objects are specific, all attempt of grasping the different shapes of life in a concept can not be express the essence of life, but it means leaving out *particular* properties of each living being that define.

For example, Lamarck and Darwin had already understood that concepts like "breed", "species" etc are nothing more than a good expedient to talk about living beings, and that these concepts do not exist in nature.[6] This means that each living being has an own irreducible specificity and, hence, a theory of biology should be *able* to explain and protect this specificity,[7] namely the fact that differences are not completely defined in a concept.

In this sense, in the following pages will be necessary to stress this specificity and understand it from the philosophical point of view. It will be very important to turn explicit the passage from possibility to reality. In this sense, distinction among possibility, potential or possible reality and actual reality, shown in relation to the quantum mechanics, will be used here. From the philosophical point of view, there are unfortunately neither satisfactory theoretical works, nor applications to biology. Let's try to trace a theoretical path that surely will have to be deepened.

From Possibility to Reality in Biology: The Concept of Thickness

Now, it is necessary to clarify the notion of "thickness of possibility". It will be analyzed from the perspective of relational space; in the next part, it will be discussed from time and organization point of view. This concept allows to take a step ahead from biological possibility to biological reality, as characterized by contingency. It has been shown that in quantum mechanics, contrarily to classical physics, possibility is separated from reality. Indeed, in quantum mechanics, not all the possible status are *able* to become real. In this sense, it is possible to conceive decoherence as a transition from quantum coherent possibility to potential or possible reality, i.e. *able* to become it *actual*. Considering the Schrödinger's cat, what has been said opens the possibility to conceive the transition from the cat as alive *or* dead ("or" inclusive, as the meaning of "vel" in Latin) to the cat as alive or dead ("or" exceptive, as the meaning of "aut" in Latin). In the former case, it is necessary to face the realm of possibility, in the latter standing in front of the realm of potential reality.

[6] For example, all kinds of willow tree are mutually fertile (interfertile).

[7] Every conceptual generalization is necessary and useful in biology in order to understand life, however it fails to capture its essence.

The measurement, namely the opening of box, corresponds to the last transition to actual reality.[8]

Conceiving decoherence as the way to step from possibility to reality, it is possible to conceive the concept of thickness of possibilities likewise to be able to step from the realm of biological possibility to the realm of reality.

In order to understand this concept, it can be useful to sketch out the relation between possibility and reality in biology. The frame is similar to the one used for quantum mechanics, but slightly modified to outline the theoretical peculiarity of biology.

Describing some biological process, for example, the reproduction (from cells to plants, to animals, etc.), it can not be forgotten that it is a matter of random process plunged into the realm of possibility because of the contingency.

From the epistemological point of view, the first distinction is between the realm of *contingent possibility* and the realm of *contingent reality*. The latter domain is divided in two: *potential* (or *possible*) *reality* and *actual* (*hic et nunc*) *reality*. It is well evident that this distinction is similar to the one in quantum mechanics. As already said, it is about a requirement that biology also demands, but differently. This difference is immediately evident considering that the list of biological possibilities is not complete and that possibility is even more strictly linked to contingency. In this sense, in the *realm of possibility*, *historical contingency*[9] of a process (mitosis or reproduction of fish, for example) allows to foresee the *set* of possible predictable phenotypes. What is being faced here is an *undetermined set* of unpredictable phenotypes. The irreducibility of biological contingency imposes to consider historical contingency, an a posteriori knowledge.

Resuming, biology is in the realm of possibility, characterized by an historical contingency; the sets of possibilities are *not* "*in general*", but are well grounded to historical contingency of processes. This implies that possibilities are always *contingent possibilities*. The transition to potential reality occurs by the *measurement* of contingent stability of environment (space). This allows to associate thicknesses with the two sets of possibilities. For example, based on an a posteriori knowledge, within a *well-fixed* environment in relation to its own history, thickness of possibility of producing predictable phenotypes is higher compared to the one of hopefulmonster. On the contrary, if stability is weak in relation to contingent history, thickness of unpredictable phenotypes is higher.[10] It results that hopefulmonster can not be predicted, neither in quality nor in quantity, but only thanks to the introduction of relational space, it is possible to measure a thickness

[8] One could clearly replace the pair possible reality-actual reality with the one virtuality-reality, but the argument should not change.

[9] "Historical contingency" is the set of data collected over multiple iterations of the same biological process, starting from well-fixed conditions. Considering reproduction, historical contingency canalizes the production of phenotypes.

[10] It should be pointed that any biological process has its own measurements in order to determine the stability of its environment, or better of relations characterizing a specific and relational biological space.

of its *possibility*. Indeed, it is important to underline that hopefulmonster is always possible and always able to become real. Possible phenotypes, characterized by a thickness, are hence able to stay in possible reality.

The transition to *actual reality* is characterized by presence or by absence (*hic et nunc*) of hopefulmonster and/or of predictable phenotypes.

It is evident that biological complexity is different from the one of quantum mechanics, because in biology the totality of possibilities is not given a priori, and because contingency plays a different role in quantum mechanics. In biology, contingency is implied also in the *creation* of possibilities.

The thickness of possibility is strictly linked to the context's contingency of biological relations. The *thickness of possibilities* could be defined as the *measurement* of influence of contingency (*hic et nunc*) on the *sets* of biological possibilities (predictable and unpredictable) such as the latter become *able* to enter in the realm of reality. Indeed, contingency of biological relations canalizes possibilities, giving them a characterization for their transition to reality. In particular, the specific configuration of environment gives the measurements of the stability of biological space. For example, during the mitosis, constraints charge possibilities with a specific thickness. In this sense, they canalize biological possibility. Its elements specify themselves and can be considered real and *able* to become actual. Thickness can not be other than *specific* because it depends on the stability of a contingent context, in relation to its passed *history*.[11]

In this sense, taking into consideration cancer cells, for example, they will have a different behavior in relation to its contingent context or to relations characterizing those cells. Soto and Sonnenschein show that, on the basis of orthodox biology, "the implicit message of the somatic mutation theory of carcinogenesis is 'Once a cancer cell, always a cancer cell " (Sonnenschein and Soto 2007, p. 125). It is important to underline in this theoretical frame that relations (in this case cellular relations) have no epistemological status. Thus, "for those who adopt the somatic mutation theory to study carcinogenesis, cellular/subcellular events conducted *in culture* provide a legitimate experimental device, since histoarchitecture (tissue organization) does not play a central role in this theory" (Sonnenschein and Soto 2007, p. 27).

On the contrary, contingency of biological context is at the core of biology, because it is not necessary, or even true, that cells behavior is always the same in any context (Sonnenschein and Soto 2007, *passim*).

> In tissue organization field theory [Soto and Sonnenschein], [...] when dissociated, cells are freed from organismal restrictions that compel them to express a phenotype appropriate to their position. Hence, they will exercise their built-in capacity to proliferate, and they may express new phenotypes, including their constitutive ability to migrate (Sonnenschein and Soto 2007, p. 139).

As for the concept of thickness, the laboratory and theoretical works of Soto and Sonnenschein show that it is possible to measure thickness of possibility, linking to relational space of biological elements and to its configuration. In this theoretical

[11] A good example is the specialization of cells during embryonic development (Villoutreix 2015).

system, it is indeed possible to *conceive* cancer regression (Sonnenschein and Soto 2007).[12]

The contingency of context is still fundamental because it allows to keep open the possibility of hopefulmonster and to apply a specific thickness to possibilities. What has just been stated is necessary as biology is conceived in the realm of possibility and not in the realm of necessity. In biology, there is *never* the totality of conditions as a biological process is *necessarily* able to only produce all the phenotypes predictable a priori.[13]

Talking about possibility in relation to hopefulmonster or to predictable phenotypes, it can not be forgotten that it is a matter of the two *sets* of possibilities and not directly to their objects, in reason of relational perspective. At this point, it is necessary to clarify how the role of biological contingency is conceived.

Contingency and Possibility

In a relational biology, the relation between contingency and possibility is very important because it characterizes how conceiving radical variation. It is a matter of conceiving *possibility as exceeding* determination *and* the fact that it is, at the same time, linked to its contingent context. In fact, given an epistemological status for relations, it is necessary to conceive possibility and the fact that biological space (theoretical and experimental) takes shape starting from the installation of relations.[14]

The shaped theoretical frame leads to a new concept of contingency, properly biological. Life is not more conceived as a general concept,[15] because it is necessary here to deal with specificity. In this sense, biology would need to conceive only what can be called the "being-there" of *singularity* of life. It is not, for example, about life conditions of a species, but about the singularity of each living being in a specific context. In other terms, contingency as "being-there".

[12] At this regards, see the next part of this book.

[13] This is a rephrase of an excerpt from Wittgenstein, although he refers to a different context (Wittgenstein 1986, §183).

[14] It has been shown that quantum mechanics explains how the elements of a system interact among each other and, consequently, how it is possible to talk about reality only within and starting from relations. In biology, relations are important not only with respect to reality, but to possibility as well.

[15] It is important to stress that all general or univocal concept of anything does not capture an essence, but it does not consider the singularity of each living being.

Contingency and Context

In biology, context is a fundamental component, which has always to be taken into consideration. Conceiving biological space as relational, as an equilibrium of tensions, context shows always itself only as specific. Indeed, relations are biological space and are never identical. In this sense, in order to be able to increase the thickness of predicting, it is necessary to define experimental context, even if hopefulmonster keeps always a possibility.

In order to show that biological context is specific or contingent, it is enough to ponder, briefly, on historically predictable possibilities of the incomplete list of a biological process. In fact, even though context is definitely well fixed, the percentages of different expected phenotypes, effectively produced during several reproductions, are *always* different for each reproduction (Villoutreix 2015).

This means that, in biology, it is not only its object that is *specific*, but also its *context* (or space or equilibrium of tensions). This kind of specificity can be explained with the idea of relational space and starting from the fact that relations have an own epistemological status. For biology, again, the same experimental contexts are *never identical*; they are *always* "*similar*".[16] The similarity of contexts or of spaces allows to compare situations not too different and historically known. Similarity of equilibrium of tensions is, indeed, fundamental to shade light onto something of biological processes. On the contrary, identity allows to conceive biological context and relations deterministically.

In other terms, there are always tiny variations produced by relations, that "self-modify" and modify their elements. Certainly, these modifications can (or can not) produce some consequences at phenotype level; nevertheless it remains impossible to perfectly reproduce two identical contexts.[17]

Contingency and historical (or a posteriori) similarity of a theoretical space or of an experimental biological context make possible to obtain measurements on biological random processes and to understand *canalization* of possibilities. In fact, it is impossible to conceive any biological process, not *situated* within a specific space. Therefore, contexts are always structured through constraints. The peculiar equilibrium among tensions canalizes biological possibilities. In this sense, constraints play a fundamental role, but they are not other than elements of biological equilibria, results of relations.[18]

From the theoretical point of view, biological interactions are never so "pure" to be released from its context. If relations, its elements and space modify themselves mutually and continuously, it is precisely contingency that allows to understand biological processes.

---

[16] "Similar" does not mean almost equal, but essentially different.

[17] Even though it was possible to reach identity, quantum physics and mechanics after Poincaré have shown that one must go beyond the epistemology of eighteenth and nineteenth century.

[18] The role and function of constraints should be better elaborated, however in this text it will be enough to show their topological place within the domain of relational biology (about constraints, see the next part of this book).

The contingency of context imposes a limitation to particular biological situations, but this also prevents to introduce necessity in biology: the situation, for which biological processes are considered as interactions always identical, will never be faced here (Buiatti and Longo 2013). Even with the hypothesis of knowing the *totality* of conditions of a process, it is not possible to foresee *necessarily* all possibilities. In fact, relations have its own proper epistemological status, beyond reductionism and determinism, even if they remain well-grounded to a particular context.

Contingency and "Exceeding" Possibility

Considering radical variation as one of main element of biological theoretical frame, it is possible to preserve a specific possibility of biology. It is about possibility of "*exceeding*", of phenotypes, whose its possibility is always present and always unpredictable (hopefulmonster). The relation between randomness and probability, relational space and symmetry breakings represent some attempts to conceive biological unpredictability. From the philosophical point of view, they have received a great importance, since it is impossible to foresee the *complete* list of phenotypes, as it leads to beyond determination and beyond *linear* causality. It is possible to affirm with Wittgenstein that:

> What is insidious about the causal approach is that it leads one to say: "Of course, a that's how it has to happen". Whereas one ought to say: It may have happened like that, and in many other ways (Wittgenstein 2001, p. 77).

It is getting clear how important is to conceive biology differently, starting from its intrinsic dynamicity and from possibility.

Indeed, all scientific tools can only build closed concepts, containing invariants of life, neglecting differences. Based on classical physics, traditional biology has tried to find an invariant with the aim to determine a priori the complete list of biological possibilities, reducing them to possible realities. Relational biology represent the effort to conceive radical variation, maintaining open the door of possibility. At this point, it is necessary to agree on the idea that biological relations and possibility of hopefulmonster would exceed the language of the science based on invariants (so it is said "static"), on determinism and on necessity. It does not mean that biology loses sense, but that it is necessary to bridge the gap and the lack on conceiving a *new* science of *life*. The philosophical experience of biological impossibility and the impossibility of predictability, already shown, can open new perspectives to conceive biology.[19] From the theoretical point of view, the epistemological elements introduced so far, constitute a coherent description to conceive the radical variation.

---

[19] The history of science has clearly shown that "résultats négatifs" ("negative result" as Poincaré named his non-solution of the problem of three bodies) can open new domains for knowledge (Longo and Montévil 2013).

In this sense, contingency is a fundamental aspect. The specificity of biological space establishes a *limit that opens possibilities* of going beyond the determination. Moreover, it explains the fact that biological variations are essential for life as unpredictable. In fact, an exceeding possibility is conceivable thanks to its relation, to its contingency and to its specific space. This kind of possibility is conceived starting from a contingent space, even if it goes beyond. From the philosophical point of view, relations refer to the breaking of a contingent determination and *show* (not "determine") that there are indeterminable possibilities, as life exceeds knowledge. In conjunction with the set of possibilities known a priori, relations, *as they are active* (*played*), open at an impossibility, neither absolute, nor abstract, but linked to a contingent space, which is relational. In this sense, relations are not causes, they *are* space that *enables* and opens possibilities[20] (Longo and Montévil 2013).

The fact that variation represents Longo et al. (2012) an invariant means that, talking about biological phenomena, it is necessary to take account of contingency, of a fundamental unpredictability. *At the same time*, contingency and the "similarity" of many biological contexts also allows canalizing possibility through the constraints, without forgetting the difference between similarity and identity. In other words, without forgetting the possibility of hopefulmonster, of the fact that, for example, randomness plays a fundamental role in biology, not only in choice of phenotypes.

From the gnoseological point of view, contingency is therefore a fundamental element as it allows to conceive also unpredictable possibility and canalization. In other terms, contingency allows the possibility as *exceeding*, but always as related to a priori determinable possibilities, since it still is in a specific context. With the same meaning, the contingency creates the necessity to conceive the canalization of possibilities, without falling within determinism (Laplace and Monod).

In other terms, placing contingency to the core of biology avoids that possibility falls into a complete indeterminacy and into determinism. It is about properties conceivable only starting from a relational structure of biological space and from the fact that biological relations have an own epistemological status.

Thickness, Symmetrization and Variation

After talking about contingency and its consequences, the concept of thickness can be better based on mathematical and biological context. The Soto and Sonnenschein's approach has already been introduced, however it is necessary to propose different elements.

[20] In this respect, they show breaking of symmetries.

What has been proposed about thickness was necessary in order to conceive biological possibility as intrinsically contingent. In this sense, it is deeply grounded in a historical contingency. Moreover, biological possibility is contingent also because of immanence of a particular context (or better of a particular space) where biological processes occur. The limitation of space and historical contingency allow to conceive possibility as *exceeding*.

From the philosophical point of view, relational biology has been characterized by means of contingency, in the sense of constitution of a theoretical space for radical variation. It is very important to strengthen such topic by a short confrontation on biological measurement (Montévil 2011). In fact, he exposes two different ways of conceiving and interpreting the biological measurement: symmetrization and the observation of *symmetries breakings*. These are two *complementary* perspectives to take into account the same matter. As to the first one, it can be stated that contingency really allow to conceive confrontations among several *historically similar* situations and to grasp elements that do not change much; in other words, it allows construction of symmetries starting from a specific history of symmetries breakings. In the context of his interpretation of biological measurement, Montévil states: "More precisely, a measurement is always performed on a specific biological object, having a specific history of symmetry changes, by taking phylogeny into account, up to a certain point, it is possible therefore to have a shared history of symmetry changes and as a result a restricted genericity of objects, defined by this common symmetry changes"[21] (Montévil 2011, p. 166). In other words, historical contingency imposes a *contingent possibility*, a particular set of predictable phenotypes. Starting from this set, if an additional process has the same initial conditions, it is possible to apply a *thickness not to this latter, but to their possibility*, in a new experimental context. Now, it is possible to understand the reason why thickness is the *measurement* of influence of contingency on the two sets of biological possibility. In this sense, symmetrization shows its connection with what has been called here "historical contingency" and it represents a fundamental *condition* in order to use the concept of thickness.

It is very interesting to underline that Montévil puts forward a number of examples that help to understand how it is possible to control experimental space in order to re-produce historically determined results. In this sense, he emphasizes the "control of the environment and the ruling out of specimens with unwanted symmetry changes " (Montévil 2011, p. 166). In fact, , *starting from its history*, fixing environment does not imply *directly* a control of possibilities, but that it is possible to apply a thickness to spaces of possibilities; it becomes viable to characterize *indirectly* production of possibilities. Obviously, it is not necessary to repeat here the ideas of Montévil, but it is very important to underline that *mathematical* symmetrization and *philosophical* thickness are just one face of contingency.

[21] The connection between "restricted genericity" and "universal limited validity" is really evident.

Contingency and limitation (only speaking about contingent possibility and reality) that it entails, allows to conceive possibility as *exceeding*: this is namely the other face of contingency. In this regard, it is enough to refer to symmetries breakings and to radical variation as the basic elements of a relational biology. From the mathematical and biological point of view, it can be said that "this approach allows then to reveal how various quantities of the control parameters leads to a change in the reorganization of the matrix by the cell and more generally to different morphogenesis" (Montévil 2011, p. 169). To sum up, contingency or "biological specificity does not necessarily leads to a fragmentation of the theoretical significance of (all) experiments. On the contrary, the creation of observables, for example, the basal metabolic rate, allows evidentiating/constituting symmetries ... and symmetry changes " (Montévil 2011, p. 170).

Starting from what has been said up to here, it is possible to find several examples from experimental biology within Montévil's text.

Constructing a relational biology and a new philosophy of biology, it should be taken back, on the one hand, to the *philosophical concept* of thickness and to the *mathematical tool* of symmetrization, and, on the other hand, to the *philosophical concept* of radical variation and to the *mathematical* symmetries breakings. It is necessary to construct a new biology starting from the concept of contingency. This means renouncing to generality and universality, typical of mathematics or physics. Accepting the idea that biological trajectories are generic (possible paths) and that biological objects are specific, it is only possible to progress in direction of contingency, proposing concept of contingency as the theoretic *institutionalization* of what is really each biological process. Here, it is important to emphasize that a "contingent" biology is not being proposed, but a "relational" one. The importance of the concept of relation and its applications to biological space have already shown, but it is necessary to examine in depth the idea of relation and its consequences.

For what concerns the link between the concept of relation and that of contingency, relations allow that contingency (for example, in an experimental context) organizes[22] itself, always differently. In fact, what has been considered here is contingent biological space and this is conceived starting from relations.

Differences Without Concept

The two exposed aspects of biological contingency impede to *generalize* the theoretical knowledge *necessarily*. What is being faced here are exceeding (undetermined) possibilities and irreducible contingency of biological context. In this sense, through relational space it is possible to conceive a biological process without *directly*

[22] It is important to remind that it can find this idea by Noble's works. He always emphasizes that relations are on the basis of biology.

talking about possibilities and foreseeing them completely a priori. Now, it is necessary to clarify what has just been affirmed. From the theoretical point of view, talking about experimental contexts, *similar* but never identical, it is impossible to conceive results of a same biological process as mass products. In this sense, it is *possible* that two similar contexts would produce same phenotypes, but never necessarily.[23] Therefore, it is correct to generalize and *partially* to foresee the list of possibilities, starting from thickness of an aleatory biological process. In this sense, biological generalizations do not imply that a biological process *should* develop pre-determinedly and that produce only expected phenotypes necessarily. It is properly its *history* (its a posteriori knowledge) that characterizes its possibilities, always in relation to its spaces and never directly to phenotypes. From the Kantian perspective, in biology there are only synthetic knowledge a posteriori (Kant 1996). Therefore, *it is possible generalizing, but never necessarily*. It is not a case that biology is neither a deterministic science nor properly a probabilistic science: radical variation adds itself unpredictably to the list of possibilities known a priori. In other terms, it is necessary to conceive biological indeterminacy at level of reality *and above all* at level of possibility. In this sense, within a same process, historically well known, there is always possibility of variations: before a bifurcation the direction of a biological trajectory is never established a priori.

In this sense, the possibility of hopefulmonster is always present, even if its thickness is feeble. In fact, it is necessary to stress that it is about the realm of biological *possibility*, if hopefulmonster is possible in a relational space. Instead, it is about the transition to the realm of *potential reality*, talking about *thickness* of hopefulmonster.[24] In fact, possibility does not depends on thickness. On the contrary, the contingency of a particular context introduces the realm of the potential reality, while thickness allows to apply measurements on the two sets of possibilities, in connection to contingent relations, implied in a particular process. The fundamental idea of relational space emerges as fundamental. Therefore, it is possible to talk about and associate measurements to sets of the possibilities, *even though their list is incomplete*.

Finally, it becomes possible to understand the title of this paragraph: it is very difficult to conceive biology without the concepts of contingency, specificity, etc. This underlines the impossibility (not only theoretical) to conceive a fundamental symmetry of biological systems.

From the philosophical point of view, the lack of necessity derives from the fact that biology answers the question "*how is*" and not the question "*what is*", more radically compared to the quantum mechanics. Indeed, answering the second question means and imposes to try to define and determine nature as something

---

[23] These are not identical iterations. In fact, if the same phenotypes are generated, percentages are different.

[24] In this respect, the biological possibility is different from possible reality of classical physics. Biological relation between randomness and probability is configured differently in physics and the biological one is irreducible to this latter.

static or well fixed. In order to answer the "what is" question, it is necessary to try and to find a definition that captures the essence of life. It is the case of DNA and the theoretical necessity of orthodox biology to conceive variations (wherever this is possible) taking them within the limits of a fundamental unity. In other terms, "what is" precedes "how is".

On the contrary, "how is" question means opening up to history, or better, to histories. Knowledge becomes historical and, consequently, it becomes impossible to get essence.[25] It is known that biological processes are random and that it is impossible to foresee phenotypes changes. Therefore, each living being has an own ontogenetic and phylogenetic trajectory, always a *possible* path, still never necessary. In this sense, in order to sum up and compare the philosophical approach of traditional biology and the relational biology's approach, the former answers firstly to "what is" and, it subordinates "how is". This means that differences are so included in a fundamental concept (or invariant).[26]

> The rigidity of the program thus varies according to the operations. Certain instructions are carried out literally. Others are expressed by capacities or potentialities. However, in the end the program itself determines its degree of flexibility and the range of possible variations (Jacob 1970, p. 10).

In a theoretical deterministic and reductionist frame, production of a living being is expression of a possibility, determined a priori, i.e. the expression of a *difference* included *in* a wider identity: DNA.

On the contrary, in a relational biology concepts are expressions of differences: "how is" precedes "what is". Concepts are contingent and, using the words of Dilthey, they can arrive only to an "universal limited validity"[27] (Dilthey 1992). In this sense, what biology can explain is valid until a variation has functional and/or morphological consequences.

Introducing the radical variation, inscribed within the theoretical frame, it is possible to conceive the expression of the differences without any fundamental identity. In other terms, it is possible "to set free" possibility from any determination a priori. Putting "how is" before "what is", concepts are determined by the elements of a system *because they interact*. In this sense, concepts are contingent and generalizable products not at all necessarily. They are not anymore expression of any essence. In this philosophical approach, concepts do not lay foundations. Their

[25] It is important to say that essence and history can be conciliated starting from the idea of an identity (invariant), showing itself in different forms during time. This option is not so interesting because it remains in a metaphysical framework. On the contrary, relational biology is characterized by radical variation.

[26] It has been shown that radical variation occurs in traditional framework, staying out from theory.

[27] It is clear that the problem of Dilthey is not biology but, from the philosophical perspective, his ideas are very interesting. He opens his essay, entitled "*Die Entstehung der Hermeneutik*" (*The rise of hermeneutics*), by posing the following question: "is it possible a scientific knowledge of singularity?". Obviously, it is impossible to follow his arguments, but it should be interesting to deepen further on the concept of "universal limited validity".

gnoseological institutionalization is valid only as historically legitimated by well-fixed relations.

> Remember that we sometimes demand definitions for the sake not of their content, but of their form. Our requirement is an architectural one; the definition a kind of ornamental coping that supports nothing (Wittgenstein 1986, § 217)

Starting from the same initial conditions it is not possible to spuriously presume that the same results will be obtained, because possibility is never constraint *necessarily*[28] to a preconceived spectrum.[29]

Incidentally, it is necessary to clarify that all the approaches that reach opposite conclusions have the task of explaining determinism, biological necessity and variation, if they want to be philosophically and scientifically coherent. At this point, it is clear that radical variation, inscribed *within* the theoretical frame of biology, imposes a likewise radical change. Unlike the Kantian "I", biology is multi-colored. In fact, in this gnoseologic and biological approach, a fundamental symmetry is abandoned because nothing represents an essential unity.[30] It is important to take into consideration the unities as *singular* and as products, in order to preserve the singularity of life and the *possibility* of radical variation.

Biology, Differences and Colors

From the strictly philosophical perspective, no bibliography exists about the subject of this essay, as it is evidently quite new. Nevertheless, some philosophical concepts can be used in order to help to conceive relational biology. "Differences without concept" has been already used by Deleuze (2000), but elements of Plato and Wittgenstein philosophy will be taken into consideration firstly.[31] It is about conceiving properly *biological* contingency and discussing relational biology through differences without concept; it is probably the solely way to conceive the fact that biological multiplicity is never something given or fixed once for ever—as Wittgenstein states (Wittgenstein 1986, §23).

---

[28] "But here we must be on guard against thinking that there is some totality of conditions corresponding to the nature of each case (e.g. for a person's walking) so that, as it were, he could not but walk if they were all fulfilled" (Wittgenstein 1986, § 183).

[29] In this sense, is has been shown sensitivity to initial conditions, the epistemological problems posed by quantum mechanics, non-linear causality, etc.

[30] Kantian analogy is clearly only an image to simplify this argument. However, it is very interesting because, in a completely different context, Kant claims that original aperception constitutes an unity, not derived from categories. Conversely these latter ones presuppose the unity of aperception. His aperception allows the "I" to not be "multi-colored" (Kant 1996, §§15–17). On the contrary, biology does not need any original or transcendental unity. Relational approach attempts to place radical variation at the core of theoretical framework.

[31] Wittgenstein expresses this idea by saying that "There is no such thing as phenomenology, but there are indeed phenomenological problems" (Wittgenstein 1978, I §53).

Indeed, the notion of "differences without concept" can be used to conceive symmetry breakings from the philosophical perspective. Moreover, it is possible to find in Plato's thought a fundamental discussion, really interesting, on the relation between unity and multiplicity. In particular, referring to the dialogue "*Parmenides*", it is evidently impossible to present here Plato's thought on this subject, but there are some fundamental aspects that he exposes synthetically and clearly. Plato's aim is different from the one of relational biology, but he discusses the pure theoretical question related to possibility that "if the one is not, in which conditions the others need to be?" (Platone 1998, 164b4-165e1). Plato underlines that if the one is not ontologically, the others (elements of multiplicity) can not be considered individually in themselves. Therefore, it is necessary that they are in relation *between them* "reciprocally" (Platone 1998, p. 164c6). This is the only way of not falling into nothing. If a fundamental unity is missing, elements should be conceived relationally. According to Plato, each element can only "*appear* one", but each of them is in fact multiple in itself. He states that they have to be conceived as "ghosts", because their ontological link to the "one" is missing. Plato's conclusions will not be followed here. Anyway, it is important to underline that without a fundamental unity (symmetry), a coherence among elements of a system could be built starting from their relations, as they usually interact. In this sense, Plato gives a general perspective without the concepts of subject and object, in a post-Cartesian sense. This is an essential point because it would be dangerous to ponder on theoretical biology introducing any intentional subject.[32] The subject's issue, very important for some interpretation of quantum mechanics, can be conceived differently by a genuinely relational perspective, like Rovelli (Rovelli 2010, 1996). Let's talk, then, rather about interactions among two or more systems instead of talking about a "subject". The concepts of subject and object are just theoretical constructions. They establish themselves into knowledge, and they are not necessarily presupposed in general or in a specific model of knowledge. The gnoseology of physics before Poincaré and the one of orthodox biology represent an example in which a kind of theoretical frame is built starting from different scientific criteria. This allows to recognize an object as object of a specific theory and to conceive a "subject" not preventing an objective knowledge. As it has already been said earlier, quantum physics introduces decisively a new role of subject within scientific knowledge. Nevertheless, Rovelli relational hypothesis should be pondered about, because it leads to a new *gnoseological* frame. For what concerns biology, a situation alike is found here. Longo's essays and the problematic frame that has been depicted show

[32] It is difficult to conceive mitosis or radical variation starting from this latter approach. In biology, it is possible to talk about well contextualized functionalities, but not about finality nor about intentionality. Moreover, evolution does not assumes any goal. The history of life is a possible history, not a necessary one. Past does not allow to predict future because it is a question of making/letting biological relations "to act" at every level. In this sens, relational framework and the idea of breakings of symmetry are sufficient for providing a consistent gnoseological approach. Shortly, if relations have a fundamental epistemological role, certain radical and theoretical changes are necessary.

that conceiving biology starting from the position or from the opposition between subject and object is a misleading, because they are theoretical constructions. Therefore, a relational approach tries to conceive biology from the dynamic point of view. In fact, elements of a biological system modify mutually, modify its space and this latter modifies *continuously* its elements.

At this point, it would be necessary to put a clarification on these new thoughts as they could appear unusual. They can be better understood through some aspects from Wittgenstein's philosophy. In his thought, Wittgenstein develops a very interesting critique about the necessity of a fundamental unity. In this sense, he thematised what Deleuze called "differences without concept" and discusses it generally through a discussion of a brief quotation of Plato's *Parmenides*.

With the aim to show what is the exceeding of a "form of life" related to the language, affirming that "*de facto*" *legitimates* "*de iure*" and that the latter *put foundations of nothing*, Wittgenstein suggests a very interesting example about colors.

Indeed, one of the aims of Wittgenstein's philosophy is unravelling the trouble of "what is" question. It has been seen that it is not necessary to assume that a definition is the end and the ultimate aim to knowledge. Concepts can be easily conceived as contingent results of an a posteriori knowledge; for example, those of biology. Whenever a concept reaches the knowledge of something, this does not express an essence, but a difference. In this sense, *preserving differences* means preserving possibility and, more than anything, its openness. This is utterly important for biology. For what concerns colors, Wittgenstein states:

> the various color concepts are certainly closely related to one another, the various 'color words' have a related use, but there are, on the other hand, all kinds of differences (Wittgenstein 1978, III § 75). The indefiniteness in the concept of color lies, above all, in the indefiniteness of the concept of the sameness of col-ours, i.e. of the method of comparing colors (Wittgenstein 1978, III § 78).

These quotations do not want to pose an abstract necessity to justify indeterminacy, reducing it to an originary determination.[33] On the contrary, Wittgenstein claims that this represents an asset, which allows thinking differently.

> If there were a theory of color harmony, perhaps it would begin by dividing the colors into groups and forbidding certain mixtures or combinations and allowing others. And, as in harmony, its rules would be given no justification [begründen] (Wittgenstein 1978, I § 74).
>
> We do not want to find a theory of color (neither a physiological nor a psychological one), but rather the logic of color concepts. And this accomplishes that which people have often unjustly excepted from a theory (Wittgenstein 1978, III § 188).
>
> The difficulties which we encounter when we reflect about the nature of colors [...] are contained in the fact that we have not one but several related concepts of the sameness of colors (Wittgenstein 1978, III § 251).
>
> There is no such thing as the pure color concept (Wittgenstein 1978, I § 73).

[33] The problem of "origin" or "originary", in a metaphysical or phenomenological sense, has no quoteizenship in the domain of biology. Actually, the concept of relation and symmetry breakings dissolve it: it returns to be inconsistent.

The mere fact that *the* concept of color does not exist, as a well-defined essence, does not constitute a problem. Indeed, it is possible to build various "logics" of colors, various ways to *keep them together* and *make them interacting to each other*. In other terms, it is about a philosophical approach of "how is" and of the fact that relations are beyond reductionism.

The choice of a specific logic will depend on use of colors in a specific context. *De facto* legitimates a logic that will be, so to speak, *institutionalized*. This latter is a contingent result and it is never a matter of foundation.

Starting from what has just been said, scientific theories *do not found* knowledge of reality, but they are only *institutionalization* and *formalization* of different ways to approach natural phenomena. Following this path, there is a "praxis" that belongs to no theory and, therefore, can not be *completely* understood into a theory. On the contrary, there is a "praxis" which *establishes* a meaning able to be institutionalized. Then, a concept is a selection of specific aspects of a phenomenon, but never its essence.

> If it is a calculation, we adopt it is a calculation—that is, we make *a rule* of it. [...]. It gives us a method of describing experiments, by saying they derivate from this by so much. [...]. If we call it a calculation, it is a complete picture which now serves as a standard or phraseology for the description of an experiment. We might have adopted $2+2=4$ because two balls and two balls equilibria four. But now we adopt it, it is aloof from experiments—it is petrified (Wittgenstein 1989, p. 98)
>
> Tractatus logico-philosophicus, 4.5: 'The general form of proposition is: This is how things are.'—That is the kind of proposition that one repeats to oneself countless times. One thinks that one is tracing the outline of the thing's nature over and over again, and one is merely tracing round the framework through which we look at it (Wittgenstein 1986, § 114).
>
> We predicate of the thing what lies in the method of representing it. Impressed by the possibility of a comparison, we think we are perceiving a state of affairs of the highest generality (Wittgenstein 1986, § 104).

In this sense, different perspectives does not imply any qualitative difference. In this framework, the idea that is not possible to characterize the scientifiquotey uniquely should be inscribed. In fact, it is possible to fill the proposition "mathematics is a language of the nature" variously. In general, it can be said that causality, reductionism, relational biology and so on, are not other than different ways to keep together some elements of reality, referring to different theoretical needs and procedures.

> 12, 14, 13, ... This would be immensely impractical, inconvenient - but not wrong. Suppose I always left out 13 in my mathematics. You might say - (a) this it is useless; (b) that it is uninteresting. And under normal circumstances it would be. But if there were people who were terrified of the number 13 this mathematics might be of great importance of them (Wittgenstein 1989, p. 83)

The different ways of keeping together various elements of reality constitute one or more spaces of possibility. In this sense, it is not about concentrating on an univocal definition or on a recurring definition, but on the spaces of possibility that can be built. It is not a matter of catching the essence of nature, but it is about pertinence and theoretical coherence. This does not mean that it is necessary to refer meaning to a praxis (for example, mathematical and scientific), but that it is necessary to conceive the meaning itself as a construction sprung up from the spaces of possibility, especially considering a discipline deeply delved into possibility as biology.

Generally, each way of *keeping together* colors is not other than a *construction of its own theoretical object* and of its significance. In other terms, they constitute spaces of possibility, but starting from interactions.

It does not emerge only the problem of plurality of perspectives, but also the question of how to conceive the realm of possibility.

From the comparison between the modern mechanics and the one after Poincaré, it does not emerge only a change about content. Non-linearity introduces new criteria of scientifiquotey. Contrarily to Lagrange, Poincaré holds that geometrical and qualitative tools are scientifically pertinent. Poincaré's maps and Lyapunov's exponent are evidently being used as reference here.

Unlike Laplace and Lagrange, starting from Poincaré, it is possible to conceive order springing from chaos. This is an example of keeping together the same elements in an unconventionally. Colors, non-linearity, quantum mechanics and biology show that it is necessary to ponder the fundamental problems of science and on how they structure themselves, *beyond an analytic perspective*.

Biology and Relations

This last chapter has been dedicated to a new philosophical interpretation of biology through the philosophical idea of relational biology. If, the first chapter has tried to show the weakness of traditional biology and its deterministic, physical and gnoseological basis, the second has shown that is utterly difficult to talk about a probabilistic biology, as aleatory and probability do not coincide at all. From this point, each aleatory *biological* process evidently implies symmetry breakings, i.e. possibility can not be thought solely *within* determination. Relational space has been proposed as the element that allows to overcome of biological impasse (not only theoretical) to reach contingency. In this sense, the double role of contingency and the fact that biology is plunged into the realm of possibility was discussed. What has just been said has another and important consequence: if the transition from possibility to reality is ruled by no necessity, but it is contingent, an *irreversibility* of biological process is faced. In this sense, Longo, Bailly and Montévil talk about "cascades of symmetry breakings" (Longo and Montévil 2014) proper of biological processes. For what concerns this thesis, its purpose is to give a theoretical and gnoseological characterization to a different biology that here has been called "relational".

The effort, at the basis of each attempt to conceive a new biology, needs to put radical variation in the theoretical frame and within all the biological processes. In this sense, it is necessary to introduce the idea of relational space and give an epistemological status to biological relations. It is a matter of creating a new form to conceive variation without subsuming it to any fundamental symmetry. On the contrary, physics proposes theories that conceive movement and variation starting from fundamental symmetry. The most important difference, not only theoretical, between physics and biology is about how to conceive the role of variation. If theoretical structure of classical physics (and deterministic biology) can admit only an epistemic indeterminacy and if quantum mechanics can admit epistemological indeterminacy too in correlation with actual reality, in biology it is necessary to consider the fact that it admits an epistemological indeterminacy in relation to *possibility* and reality. The theoretical relation, that radical variation implies when considering biological theoretical frame, has been shown here. Biological indeterminacy of possibility leads towards symmetry breakings and to the fact that biological processes are not necessary and, hence, ontogenetic and phylogenetic trajectories are not reversible. In general, biology is immersed into the realm of possibility and it is deeply characterized by contingency.

In the next part, a lot of concepts, exposed until now, will be exposed starting from a new interpretation of biological times and rhythms.

Bibliography

Bailly, Francis, and Giuseppe Longo. 2006. *Mathématiques et Sciences de la Nature*. Paris: Hermann.

Buiatti, Marcello, and Giuseppe Longo. 2013. Randomness and multi-level interactions in biology. In *Theory in Biosciences*, vol. 132, pp. 139–158. https://doi.org/10.1007/s12064-013-0179-2.

Deleuze, Gilles. 2000. *Différence et répétition*. Paris: PUF.

Dilthey, Wilhelm. 1992. *Ermeneutica e Religione*. Rusconi: Milano.

Heisenberg, Werner. 1971. *Physique et Philosophie: La Science Moderne en Révolution*. Paris: A. Michel.

Heisenberg, Werner. 2010. *La partie et le tout. le Monde de la Physique Atomique*. Paris: Flammarion.

Jacob, François. 1970. *La Logique du Vivant*. Paris: Gallimard.

Kant, Immanuel. 1996. *Critica della ragion pura*. Laterza: Roma-Bari.

Kupiec, Jean-Jaques. 2012. *L'ontophylogenèse*. Versailles: Quæ.

Longo, Giuseppe. 2010. Incompletezza. In *Per la Matematica*. Vol. 4. Torino: Einaudi.

Longo, Giuseppe, and Maël Montévil. 2013. Extended criticality, phase spaces and enablement in biology. In *Chaos, Solitons and Fractals*. Elsevier.

Longo, Giuseppe, and Maël Montévil. 2014. *Perspectives on Organism: Biological Time, Symmetries and Singularities*. Berlin: Springer.

Longo, Giuseppe, Maël Montévil, and Stuart Kauffman. 2012. No entailing laws, but enablement in the evolution of biosphere. In *GECCO Companion '12*. New York: AMC.

Monod, Jacques. 1989. *Le Hasard et la Nécessité*. Paris: France loisir.

Montévil, Maël. 2011. Temps biologique et transitions critiques étendues. Thèse de doctorat. École normale supérieure de Paris.

Omnés, Roland. 1994. *The Interpretation of Quantum Mechanics*. Princeton: Princeton University Press.

Platone. 1998. *Parmenide*. Laterza: Roma-Bari.

Rovelli, Carlo. 1996. Relational quantum mechanics. *International Journal of Theoretical Physics*. 8th ser. 35: 1637–1678.

Rovelli, Carlo .2010. *Quantum Gravity*. Cambridge: Cambridge Universit Press.

Sonnenschein, Carlos, and Ana Soto. 2007. *The Society of Cells*. New York: Taylor & Francis.
Tagliagambe, Silvano. 1991. *L'epistemologia Contemporanea*. Roma: Editori Riuniti.
Villoutreix, Paul. 2015. Randomness and variability during embryogenesis A multi-scale approach. Thèse de doctorat. École normale supérieure de Paris.
Wittgenstein, Ludwig. 1978. *Remarks on Colour*. Oxford: Basil Blackwell.
Wittgenstein, Ludwig. 1986. *Philosophical Investigations*. Oxford: Blackwell.
Wittgenstein, Ludwig. 1989. *Lectures on the Foundations of Mathematics*. Chicago/London: The University of Chicago Press.
Wittgenstein, Ludwig. 2001. *Pensieri Diversi*. Adelphi: Milano.

Part II
Biological Times and Organizations

The Problem of Time

The first part of this text dwells on the concept of space. It allows to develop an original and solid base for *relational biology*. After highlighting some fundamental problems of traditional approach to biology, new principles have been stated, starting from theoretical enhancement of symmetry breakings and their application to biology.

It has been shown that relational biology can conceive and structure coherently a mathematical approach based on symmetry breakings and that it safeguards the singularity of living beings. In particular, the field of theoretical biology (the creation of possibilities) has been defined by the concepts of enablement and thickness of possibilities and by a philosophical-epistemological discussion about the concepts of possibilities and reality in classical physics, quantum mechanics and biology. Nevertheless, some concepts, such as thickness, need further investigation, while others, such as "time" and "entropy", have not yet been discussed.

This second part is devoted to show how they can be included in the theoretical framework that help to build. Thus, it will be possible to frame typical biological concepts such as "organism", "organization", etc. within relational biology.

In relational biology, the concept of "time" represents precisely the middle term that allows to link the latter concepts mentioned to that exposed in the first part, in order to constitute the complete picture of relational biology.

As for the concept of space, that of time must also be adequately explored because biology needs an own notion of time, different from the physical one.

Variation is considered as the fundamental invariant[1] of relational biology, in contradiction with Monod, where DNA is considered as the essential invariant of a reductionist biology. It will be shown that in works of organicites too, organization recalls a fundamental invariant. It is a matter of two perspectives completely different and irreconcilable because the last two are based on a principle

[1] Obviously, it is a matter of an invariant contradicting the concept of "invariant".

A. Marinucci, *Theoretical Principles of Relational Biology*,
Human Perspectives in Health Sciences and Technology 6,
https://doi.org/10.1007/978-3-031-39374-7_4

of conservation and consequently they have to explain biological variation. On the contrary, relational biology, based on variation, has to explain the fact that a specific organization is not only formed, but also maintained. This is very important in the attempt to constitute general basis of a theory of organism, in a relational approach. In general, if one wants to keep the singularity of the living beings and a biology based on symmetry breakings, it is not definitively possible to admit organization as theoretical principle.

Preliminary about the concept of time, it can be affirmed that biology's times and rhythms have to be at the base of life and of every scientific approach. The rhythm of heart, breathing, etc. are also important from the phylogenetic and ontogenetic point of view and still in mathematics applied to biology. In fact, differential equations are always written as functions of time, normally conceived as container in which biological processes occur. This is the Newtonian image of time. In this sense, the space as container has been abandoned in order to conceive a relational space; similarly, it is also necessary to change the concept of time. This would be essential if one wants to safeguard the singularity of living beings theoretically. It has been shown that, from this perspective, it is not justified to conceive singularity as a particular case of a general invariant. On the contrary, it is necessary to conceive every general concept of biology as abstraction that, if on the one hand, allows to conceive a set of phenotypes, on the other hand, loses the singularity of each living being. Shortly, it will be necessary to radicalize the idea of Darwin and Lamarck, according to which concepts of race, species etc. do not exist in reality: it is only abstractions that allow to talk about a group of individuals, who remain always different from each other. In this sense, it is possible to understand the reason why, for example, there is no single rhythm of the heart for stay healthy. On the contrary, there are values, a minimum and a maximum, out of which it is a matter of disease, but without any necessity, because the rhythm of heart is only a result of organic processes, of organic relations. *This means that it is impossible to interpret biological beats and rhythms without considering the relational space that produces them in its own specificity*.

Biological times are results that make possible to have measurements on relational spaces or on *interactions between organism systems and subsystems*, that are the *theoretical observables of relational biology*. From the practical point of view, biological observables are all the forms in which a relational space can occur, from tissue to organism, from environment to ecosystem etc. For obvious reasons, this text follows an exclusively theoretical approach.

Generally speaking, relational approach to biology leads to the core problem: are biological times and rhythms results of organic relations or, like classical physics, are they what, in function to which and *in* which one can describe biological systems? In particular, are they results or "containers" and/or what allows to describe nature? The way to answer to these questions allows to conceive times in a relational perspective and to conceive, finally, a new biology.

In this sense, it is necessary to show briefly how to conceive the concept of "time" in general and to deepen it in relational perspective. Then, it is necessary to deepen also the concepts of "entropy", "negative entropy" and introduce the one of "anti-entropy". When this latter concept will be exposed, it will be possible to clarify relational times, the concept of thickness and, finally, to introduce the concept of organism into this philosophical and biological-theoretical structure.

In summary, the contents of this second part can be summarized as follow. Every biological times, rhythms and "harmonies" are characteristic, they respect the singularity of living beings and are subject to symmetry breakings. From the mathematical point of view, a change is not described with respect to time, but a time or a rhythm is described with respect to a change. Therefore, it is a matter of biological "time*s*" and not of physical "time". Through a brief discussion of some elements of the concept of entropy, it will be shown that it is possible to use time to establish if and how much a certain organic configuration is *stable or not*. As seen in the first part, this means determining the *thickness* of possibilities of relational space.

Therefore, through times, rhythms and thickness, it becomes possible to talk about (and reach) specific measurements on relational fields at each level (tissue, organ, apparatus, organism, etc.); in other words, it becomes possible to talk about an *organism without losing its singularity*.

The main effort will be to change the normal and intuitive way to conceive time, vitiated by physics, especially not relativistic.

Considerations About Physical and Biological Time

Before introducing different ways to conceive the concept of time, it is important to state that there is no necessity to consider time as a container or as the variable in function to which it is possible to describe movement. This is essential in science, because scientists often assume the same way to understand scientific concepts. Obviously, this allows to create semantic networks, but it prevents to think different, hence it is necessary call into question all what is "evident". In this sense, into history of thought, it is possible to find neglected or new ideas to conceive nature differently. The aim of this second part is to combine old and new ideas to conceive differently biology and organism.

A biological or physical system can be described starting from its temporal evolution or from its properties, in order to understand how it changes in function of time or in function of its properties. Focusing on symmetries and symmetry breakings, it is possible to deal directly with properties of a physical or biological system and only indirectly with its temporal evolution.[1] The main consequence is that it is possible to conceive temporal evolution of a system starting from changes of its properties; in other words, starting from specific symmetry breakings.

In a theoretical biology, whose "fundamental invariant" is *variation*, it is necessary to conceive the constitution of *specific* biological *times and rhythms* (plural), starting from what is found in nature (just consider the diversity and plurality of life rhythms, even within the same species). In this sense, physical time can not be uncritically applied to biology. It has to be conceived differently, according to preservation of specificity of biological objects, reduced frequently to the physical one.[2]

[1] Of course, it is not possible to exclude the possibility of describing a system from its temporal evolution, but the other perspective is more interesting, at least in biology.

[2] An important reference has already been made to the fact that Longo and Bailly affirm that biological object is specific, while the physical one is generic, but it is important to reiterate it

A. Marinucci, *Theoretical Principles of Relational Biology*,
Human Perspectives in Health Sciences and Technology 6,
https://doi.org/10.1007/978-3-031-39374-7_5

In physics and in history of physics, there are already many studies about this, while in biology, unfortunately, there is no sufficient bibliography. Thus, a physical and mathematical *excursus* is necessary to finally reach biology.

Measuring Time and Measuring Movement

From the sixteenth century, there was a real revolution in techniques and technology of time measurement enabling an important improvement of scientific observations (Proverbio 1983, p. 1). Of course, it is not necessary to expose here everything happened; in fact, it is enough to show that using time to describe movement, at the same time, movement is used to calibrate clocks. In his *Racconto istorico*, Viviani said that Galilei thought that it is possible to apply it to the measurement of time, after observing oscillations of a chandelier in dome of Pisa. Galilei began to use it to measure the pulsation of heart.[3] From the broader perspective, Galilei poses the problem of "time meter" (*misuratore di tempo*). In fact, in a letter to Baliani (September, 1, 1639), Galilei tells about calibration of "misuratore di tempo", saying that it is based "on the measurement of oscillations contained in the interval between two passages of a star in relation to a fixed reference (Proverbio 1983, p. 69)".

From the theoretical point of view, it can be immediately emphasized that Galilei compares two types of movement in order to improve the accuracy of its instrument. In this sense, he derives *a* time from movement and then applies it to measure the latter. Clearly, it is not an absolute time, but a time established from a movement, hypothetically always "periodic". In this sense, a measurement of time has been defined from the relationship between objects. Recalling Descartes, nature is reduced to matter and movement and, if one is able to measure time, it becomes possible to know nature completely.

In this sense, it is not difficult to understand how it was, and still is, important to calculate time as precisely as possible, but it is never to forget that time is constructed from movement.

---

especially regarding to organicistic approaches (from Varela onward) which consider organization as a principle, but without proposing an invariant. The point is that Longo and Bailly's observation digs an unbridgeable groove between biology and physics to ensure a theoretical space proper to biology. In fact, organicist approach reduces the biological object to physical one, precisely because it claims to find a general invariant.

[3] Viviani (1907, p. 603): "... with the sagacity of his ingenuity he [Galilei] invented that very simple and regulated measure of time by means of the pendulum, not previously perceived by anyone else, taking the opportunity to observe it from the motion of a lamp, when one day he was in the Cathedral of Pisa; and making very exact experiences of it, he ascertained the equality of its vibrations and at that time it occurred to him to adapt it to the use of medicine for measuring the frequency of the wrists, to the amazement and delight of the physicians of those days and as it is also normally practiced today: he made use of this invention in various experiments and measurements of times and motions, and was the first to apply it to celestial observations, with incredible purchase in astronomy and geography".

In fact, it is not necessary to describe how an atomic clock works, because, although it is technologically different from that of Galilei, it is always a matter of focusing, if not precisely on a "movement", however, on something that is not "temporal".

Therefore, it emerges that through motions, already called periodic, it is possible to synchronize the watches or, better, the movement of their pieces. As a result, movement is used to set clocks, but then these are used to measure movement. In fact, although in a very different context, Aristotle already stated that time is the measurement of movement, however, immediately after Galilei, two hypotheses were put forward to conceive the nature of time (and space).

In this sense, there are two possible choices to conceive time either absolutising it or relationally; these are the two theoretical options represented, respectively, by Newton and Leibniz (Newton 2008; Leibniz and Clarke 2007). It is easy to understand that the first option allows to conceive only one time, on the contrary, the second one allows several times. If time is conceived relationally, it is possible to safeguard the different singularities of living beings because each rhythm and biological time is a result of relational sets constituting each organism.

From the metaphysical point of view, the question is whether time or movement is "more original". From the relational point of view, the problem of origin or priority is a false problem,[4] it is important to analyze various options and decide which is the most *pertinent*, since there is no legality of nature outside what is imposed it (Kant 1996) or, more radically, outside an interpretation field (Marinucci and Crescenzi 2021).

Time from Movement

Immediately after the reception of Newton's *Principia* in France and Germany and after translation of Newtonian geometric calculus in Leibnizian symbolic calculus (Blay 1992), time becomes the variable in function to which movement is described ($\frac{dx}{dt}$) immediately and tacitly. The fundamental assumption is that *absolute* time *exists*; it is not only a mathematical element, within movement takes place and with respect to which the latter is described, time becomes the privileged variable to study natural phenomena.

This idea is questioned on basis of some features in general relativity, up to the point that Rovelli comes to argue that time is neither a fundamental element of nature nor it should be considered as the privileged element of scientific description. Actually, it should be conceived as an emerging phenomenon (Rovelli 2008, 2017). As far as relational biology, Rovelli's ideas will be partially used because, *mutatis*

[4] This issue will be clarified in the last chapter. Using a geometrical point of view (symmetry), it is possible to avoid theoretical problems as reductionism in order to focus on what emerge from the relation among elements of a system.

mutandis, they can be productively applied to theoretical biology. In fact, relational biology deals with times that are results, so to speak, of movement.

This topic will be better developed in the next chapters of this text; at the moment, just remember that Noble (2006)[5] has shown that it is precisely the whole organism contributes to heart rhythm. Therefore, heart beat can be conceived as a *result*, not as what with respect to which movement is described. After all, this new interpretation of time only deepens the fundamental difference between physics and biology, expressed by Bailly and Longo (2006) and already mentioned in this text: if in physics trajectories are specific (geodetics) and objects are generic, in biology trajectories are generic (possibilities) and objects are specific (historical-empirical). Therefore, each object has its own evolution, that it is not very accurate to call it "temporal", although it certainly remains a possibility (Longo 2016). Hence, relationships, constituting it, are never given completely *a priori*, as their results which (as far as living organisms are concerned) can be expressed through times and rhythms.

About the concept of time, new theoretical elements are developed in general relativity by Carlo Rovelli (2010), who suggests that time should not be considered as the privileged variable in physical description of nature. According to Rovelli, it is always possible to describe movement with respect to time, but also to describe it with respect to any other parameter.[6] Obviously, now it is not a matter to take a position with respect to Rovelli's ideas, which, moreover, present various philosophical problems; it is important to discuss only his perspective insofar as it is relevant to relational biology.

Indeed, at least partially, it has already been taken into account in the first part of this text with reference to the concept of relational space. The Rovelli's ideas about time are even more disruptive, even if they fit into a well-defined tradition (Leibniz-Mach-Einstein). As mentioned, in classical physics, up to Einstein, time has always played a fundamental role, as evolutions of systems were described as functions of time: "in pre-relativistic mechanics, time is a special physical quantity, whose value is measured by physical clocks, that plays the role of the independent variable of physical evolution (Rovelli 2008, p. 2)".

One of consequences of theory of relativity is that it is impossible to determine what happens to an isolated object at a specific point in space-time, because the latter is not conceived like an inert container (Newton). Hence, measurements describe directly the dynamics of space-time itself which, relationally conceived, brings to the center of physical descriptions the notion of "tensor of curvature" of gravitational field (Rovelli 2010; Lachièze-Rey 2008). Hence, its evolution does not take place with respect to time, but with respect to relational space and what constitutes it.

[5] Some ideas of Noble will be discussed specifically in the last part of this text.

[6] In this sense, the Wheeler-De Witt equation does not consider " time" as variable.

As for the concept of time, it loses its central role that it had in pre-relativistic physics because the measurement of curvature tensor depends on relations among objects that *are* and constitute gravitational fields.

If Galilei already set clocks through celestial movements, today, starting from scientific discoveries related to relativity, the so-called "flow of time", or the functioning of the clocks measuring it, depends on gravitational field. In fact, taking two clocks and putting one on Earth and one in space or near a black hole, it is well known that time passes, or rather, is measured differently by the same instrument. It is clear that it is possible to associate these various "times", but the important thing is that:

> in general relativity, there isn't a *preferred* and *observable* quantity that plays the role of independent parameter of the evolution, as there is in non-relativistic mechanics. General relativity describes *the relative evolution of observable quantities*, not the evolution of quantities as functions of a preferred one. To put it pictorially: with general relativity we have understood the Newtonian "big clock" ticking away the "true universal time" is not there (Rovelli 2008, p. 3).

Recent studies (Bizzarri and Cucina 2014; Testa et al. 2014) have shown that in absence of gravity cellular development is different from what happens on earth and this, as can be easily understood, has repercussions on times and life rhythms that can, only in a relational biology, be considered as results, described starting from organic interactions at various levels, without considering time as a "privileged" variable.

About physics, if the concept of time can be considered differently from what happens in pre-relativistic physics, unlike what Rovelli seems to affirm, relational biology is at level of interpretation and not at the ontological one. Nevertheless, it would remain to understand how to explain the perception of flow of time. In this sense, Rovelli focuses on entropy (Rovelli 2015) and on the fact that time emerges from relationships, implying movement and/or change.

As for relational biology, it is not necessary to dwell on a similar topic, however, this last argument allows to introduce, abandoning Rovelli's ideas, issues more directly related to biology such as negative entropy and anti-entropy. Before dealing with these topics it is necessary to show why it is important to conceive time starting from movement in biology. In this sense, a quote from Noble is interesting because it represents an example of biological time considered as a *result* and because it can be interpreted within relational biology.

> That, then, is how I came to take my experimental results and a few hand-waving feasibility calculations along to the guardians of the Mercury computer. I must have explained, breathlessly, and in mathematically quite naive terms, what I hoped to achieve. I would fit my data to some non-linear expressions, I thought, then would solve the differential equations for the electrical state of the cardiac cell and then, (hey, presto!), I would see oscillations emerge from the computer's output.
>
> A single question stopped me in my tracks. 'Mr Noble, where is the oscillator in your equations? What is it that you expect to drive the rhythm?'
>
> I was speechless. I had no idea how to reply.
>
> On a sheet of paper, I sketched out the physiological interactions that I thought would work. This served only to confirm my mathematical naivety. There was no oscillator

function in my equations. Thirty years later, and after many further rounds of mathematical modelling of cardiac rhythm, I was asked almost exactly the same question by the science correspondent of a national newspaper. By that stage I knew the answer: Silly question! (Of course, I did not put it quite that way to the journalist!)

Yet, it looks like a reasonable question. In a system that oscillates, it seems that there must be some specific component that oscillates, around which the behavior of the entire system is geared; and there must be a mathematical function that describes the way that component oscillates. Indeed, it is an eminently necessary question, if we are talking about some man-made, mechanical systems. But we are not. Instead, we can have a system that operates rhythmically and yet contains no specific 'oscillator' component. There is no need for one. The reason is that the rhythm is an integrative activity that emerges as a result of the interactions of a number of protein (channel) mechanisms.

So there is no need for any oscillatory equation at the molecular level. The rhythm is a systems property. Some biologists have called these properties 'emergent' properties. I prefer 'systems-level' properties, but we are talking about the same kind of phenomenon (Noble 2006, pp. 59–60).

This quote will be commented only after discussing entropy, so let's start with entropy.

The Entropy Between Physics and Biology

It has been shown that it is not at all necessary to conceive time like a container in which biological processes take place. On the contrary, it is possible to conceive physical time as an emerging phenomenon and, in this sense, emerging from relationships. While it is true that physical time interacts with biological systems, it is also true that one can not conceive physical time, particularly entropy, as the *necessary* direction of state passages of energy. Even in cases where this direction is actually found in biology, it is always a matter of studying how it is produced and what are its specific properties because time shows itself to be a result of a relational space in biology. If it is only a measurement, living systems (biological relational spaces) interact with specific physical states. In this sense, from the theoretical point of view, it is not completely correct to consider entropic time as the time *in which* biological processes take place: it is one of the most important physical constraints that is related to biological relational spaces. In fact, following Longo's ideas, it is a matter of a typically biological entropy that needs to be conceived *in* the theoretical context of relational biology: anti-entropy.

It is very important to ponder over time and physical entropy because, from the theoretical point of view, it has brought about many decisive theoretical novelties that have influenced theoretical biology.

Mathematical tool of probability is indispensable and, in the same way, statistical approach is equally well developed (Jean-Jacques Kupiec 2012), precisely because Kupiec's researches have shown fundamental and pervasive randomness of biological processes. Nevertheless, within a relational approach, probability and statistical mechanics are not able to play a central role, since it is necessary to know *a priori all* possible phenotypes, in order to give a probability. From the theoretical point of view, this is an insurmountable limit, but it does not affect the practical use of probability. In fact, it becomes important *if* it is applied to the thickness of possibilities of the two *sets* of predictable and unpredictable phenotypes and in many other cases. In particular, it is a matter of considering measurements, normally concern phenotypes, as measurements on relational spaces. Therefore, probability is a fundamental *practical mathematical* tool, even if it is not a sufficient *theoretical* basis to conceive *biological* concept of randomness. The application of statistical mechanics and its innovative tools to relational biology requires a deepening, since this is built significantly different from physics.

Entropy in Physics

Although both ergodic hypothesis and relative equiprobability of states are fundamental for statistical mechanics (Gallavotti 2010), it is necessary to emphasize the importance of large fluctuations which, although extremely rare, can occur in physical systems (Vulpiani et al. 2014b). From the theoretical point of view, they can not be excluded *a priori*, even if *H* theorem shows that Maxwell distribution is the most probable. Precisely through the analysis of results of statistical mechanics, probability has assumed its own gnoseological status; in other words, it allows to interpret some specific natural phenomena, without understanding it, like Laplace, as a distance from truth.[7] In this regard, Boltzmann states:

> The determination of average values is the task of calculating the probability. The problems of the mechanical theory of heat are therefore problems of the calculus of probability. But it would be a mistake to believe that heat theory is for this reason subject to uncertainty, because theorems of the calculus of probabilities are used there. Do not confuse a law that is proven incompletely, whose correctness as a result is problematic, with a law perfectly proved law of the calculus of probability (Boltzmann 2010b, pp. 103–104).

Mutatis mutandis, such a statement could also have been endorsed by members of the school of Copenhagen, however, a true probabilistic interpretation of *H* theorem is only given after Boltzmann's answer to Lodschmidt's objection (Cercignani 1998, pp. 97–100): based only on *reversible* laws of mechanics it is not possible to establish a *necessity* for any kind of irreversibility in evolution of any physical

[7] Laplace states: "More often than not, the phenomena of nature are complicated by extraneous causes: an enormous number of disturbing causes mix their influence with them, so much so that it is very difficult to recognize them. In order to reach it, it is necessary to multiply the observations or experiments, so that, by mutually destroying the extraneous effects, the average results highlight phenomena and their various elements" (Laplace 1840, pp. 298–299). In addition to specifying that average values are what should be considered, Laplace argues that perturbations are mere noise. Maxwell's distribution function "considers only the number of molecules that have a certain property and not the relationships (dynamics) that exist between them" (Badino 2010, p. 14). It is well known that the idea that probability has its own epistemological status is at basis of quantum mechanics.

system. However, if this necessity can not be demonstrated it can be shown that the state of thermal equilibrium (the evolution in which entropy increases) is the most probable state or, better, its basin of attraction is extremely larger than the one in which entropy decreases. In other words, oscillations are always possible and cause changes from equilibrium to non-equilibrium. The above makes at least doubtful the simplification of Cournot principle ("it is practically certain that an event with a very low probability will not happen" (Cournot 1986)) because, as mentioned, after Boltzmann's reply to Lodschmidt probability begins to have an its own epistemological status. Although rare fluctuations, having an opposite direction to entropy are on the margins of the Gaussian, they are always possible and they should be considered as such. Although large fluctuations in physics do not have a Gaussian behavior (Vulpiani et al. 2014a), as always happens in classical mechanics, its space of possibilities too is always given *a priori*, on the contrary, in biology the complete list of possibilities is known only *a posteriori*; for this reason, biological research about entropy needs a different approach. In statistical physics, possible states of a system can be reduced to the set of possible combinations of its elements.

The fact that entropy, its direction and the arrow of time are closely linked to probability, allows to affirm that, from the physical point of view, other temporal directions are possible and that it is not very accurate to consider entropy as time *in which* living beings are *without necessary clarifications*. In particular, it is necessary to understand how to reconcile physical entropy and biological relational time, since it is a result.[8]

From the physical point of view, living beings are always *open* dissipative systems. This imposes to consider its boundary conditions from time to time, because they differ by their specific contingency: it is far from obvious to apply to them the properties of entropy, because there are quantum phenomena introducing an epistemological indeterminacy in biology.

From the biological point of view, observations made so far are far from sufficient. Schrödimger (1995) began to focus on the fact that organisms equilibrium entropy by "negative entropy". However, his idea turns out to be insufficient as it keeps biology basically in the realm of physics and works with notions that today seem to be largely outdated. Nevertheless, the element that remains interesting is that living beings, although subject to entropy, are able to oppose it, always as an open dissipative system. In a physical approach, Schrödinger only cares about what is actually able to counteract the increase in entropy within organisms. Now, the ways to oppose entropy can be various and so it can be state that, from the biological point of view, every living being can produce a resultant for which times of each organism and its historical course are different from each other and from any asymptotic function. Respecting theoretical model of physics, it is possible to

[8] As it will be shown, in biology it remains possible to describe change "into" time, but, in the last instance, it is always "marked" by changes of space of phases and by rare events, where there are symmetries breakings. As in general relativity (Lachièze-Rey 2008), once a reference system is provided, it is possible to study its objects, referring to them system measurements, even if they are always linked to spaces and only indirectly to phenotypes.

state that if energy is conserved, in biology it would be possible to rely on one or more fundamental invariant principles, such as DNA or organization. However, if the first option is almost outdated, the second presents strong problems because, it is absolutely necessary to show that *a biological element* is conserved, in order to be able to pose or presuppose a "principle of organization", without slipping into chemistry or physics.

In general, Schrödinger's ideas about negative entropy appear to be too tied to physics. The aging process is undoubtedly a good example because it shows the limitation of negative entropy: it is not only a question of balancing entropy, it is a question of considering a resultant greater than zero, a real entropy production by organisms.

For this reason, it is necessary to introduce the notion of "anti-entropy" (Bailly and Longo 2009; Longo and Montévil 2012c; Chollant-Namy and Longo 2022; Montévil 2021). It allows to show even more the necessity for a properly biological treatment of time, which is not a mere re-proposing of the notion of time and entropy proper to physics. In this way, it is possible to go far beyond the mentioned text of Schrödinger, whose full title is, in fact: "What is life? The living cell from a physical point of view".

Entropy and Relational Biology

Let's start with a quote:

> This generating and re-generating activity, from embryogenesis to repair and turnover, is typically biological and it has been mathematically defined as "anti-entropy" [...]. In other words, irreversibility in biology is not only due to thermodynamic effects, related to the production of energy, typically, but also to all processes that establish and maintain biological organization. [...]. That is, biological reproduction, as morphogenesis, is *intrinsically joint to variability* and, *thus, it produces entropy also by lack of (perfect) symmetries*. By this, it induces *is proper irreversibility*, beyond thermodynamics (Longo and Montévil 2012c, pp. 3–4).

In this passage, it is argued that, in general, there is an irreversibility proper to biology. It should be added that, if these concepts are developed considering the singularity of living being, it is possible to affirm that every living being produces its *own* irreversibility. More generally, each biological object has a history of its own, very similar to that of members of its species, but not the same. This apparently trivial clarification is very important because, starting from the principles of relational biology, "*similar*" does not mean "almost equal", but *essentially different*. Biological processes constitute and maintain the organization Through production and exchange of energy, as they always take place in a specific context. Therefore, different organizations of living beings are ontophylogenetic results, whose constraints allow the formation of those *always contingent* similarities among organisms that make possible to classify living beings variously. As seen, from time to time something common is identified, but this does not move towards the "essence" of life, this moves towards *subtraction* of singularity characterizing each living being.[9]

In the cited article, reference is made to the fact that in biology the production of order and disorder are closely linked. In fact, from the general point of view, it is possible to conceive symmetry breakings as producing disorder, but, at the same time, it is necessary to produce new forms of order; indeed, it is precisely in this way that variation is conceived within a relational perspective.[10] In fact, the notion of anti-entropy allows to emphasize another aspect of relational approach, more closely linked to thickness of possibilities. Symmetry breakings produce order and disorder, increase and decrease entropy, even if algebraic sum remains positive.

In the discussed article, there are mathematical details, but since Boltzmann's entropy was mentioned, it is interesting to quote a passage on combinatorial complexity, distinct from morphological and functional ones (Longo and Montévil 2012c, pp. 11–17).

> Thus, more soundly S is defined by $S = k_b log(\Omega) - k_b log(N!) > 0$. This symmetry by permutation reduces the size of the microscopic possibility space, and, as a result, entropy. In our approach, we have $K_c = log(N!) - \sum_i log(n_i!)$ which is greater than 0, as soon as there is more than one cell type. Thus, the increase of the possibility space (the diversity or the differentiations) increases the complexity. More precisely, the complexity, as absolute value of anti-entropy, is decrease by the remaining symmetries, quantified by the term $\sum_i log(n_i!)$. We understand then that anti-entropy can be analyzed, at least in this case, as an account of how much biological symmetries are broken by the cascade of differentiations (Longo and Montévil 2012c, pp. 13–14).

If Boltzmann's formula can be interpreted as a drastic an *probably* decrease of micro-configurations, $S = k_b log(\Omega) - k_b log(N!) > 0$ represents a reduction of the space of possibilities and, therefore, it does not represent well what happens in biology. The last mathematical expression shows how order and disorder are almost inextricably linked. From the thermodynamic or form (probabilistic) biology point of view, it is about a simple combinatorial question, on the contrary, from the relational perspective relations produce phenomena that go beyond combinatoriality, *increasing* the space of biological possibilities, as it takes into account the possibility of hopefulmonster.

The fact that $K_c = log(N!) - \sum_i log(n_i!) > 0$ shows that it is a matter of a mechanism that is not cyclical, but that feeds on itself,[11] in the sense that variation, linked or not to biological entropy, affects both order and disorder.

---

[9] Replacing one invariant with another (God, DNA or organization) simply means changing content and maintain the same thought pattern. Where a new approach is proposed, it is necessary to work mainly on method.

[10] A very clear physical example is ferromagnetic transition. Its limitation is that, unlike biology, there is no "density" of critical transitions (Bailly and Longo 2006).

[11] From the medical point of view, this does not imply that such an increase is always positive for the preservation of an organism, but, from the strictly biological point of view, it certainly represents a very important element from all points of view.

From this point of view, since every living being interacts differently with environment and it is characterized by internal biological processes, it is possible to state that every living being produces its own anti-entropy. Therefore, the production of entropy is discharged both inside and outside organism producing different effects, such as aging in multicellular organisms.

Of course, a thermodynamic analysis refers to general average values. However, at every level, from cell to organism, the production of entropy is linked to the configuration of relational space. Hence, it is possible to consider all the measurements related to anti-entropy as results describing the general situation of an organism or of an organ. This aspect can be deepened by studying organization of relational space and, in particular, its complexity.[12] Times and rhythms of organism are marked by relational space and its configurations, in relation to symmetries and symmetry breakings, i.e. to the properties of each symmetry and to their changes. In this sense, in biology there is no time marked by a periodic motion of stars mentioned by Galilei, but there are rhythms showing how possibilities and configurations of each relational space are structured.[13]

Disorder, order, increase or decrease in complexity are results that can be expressed through times and rhythms of organism; hence, it is possible to build a specific line of development of an organism itself, that remains a possible and never necessary path. For this reason, it is necessary a general criterion to be able to interpret the general stability of an organism.

In the quoted article, Longo and Montévil claim that the production of entropy is linked to symmetry breakings. The direction of entropy depends on cascades of symmetry breakings, therefore, it is specific. Since these are general measurements on organs and/or organisms, (dynamical) biological observables are precisely these relations, referred to tissue (or other relational spaces) rather than to cell itself. In this sense, it should be stressed that the minimum unit for every biological investigation is tissue (Sonnenschein and Soto 2007), not cell or. More generally, this unit is the configuration of a relational space or the properties of a specific symmetry. A cell is interesting insofar as it has an effect on its tissue or on organism. The evolution of a relational space has to be described with respect to the *stability* of tissue, *not* with respect to *time*, since the latter is a result of such description. This is because, in a relational perspective, relationships *are* relational space and, moreover, symmetry breakings at cellular level *may not* imply symmetry breakings at tissue level. For this reason biological observable is not constituted by cell, but by relational space. More generally, this applies to all biological levels. For example, talking about organism, observables are all the elements that constitute it as a relational space (organs, etc.). The fact that through anti-entropy is possible to

---

[12] Fractality is certainly an extremely useful tool in this regard. From the general point of view, it shows a not absolute invariance, but a constructive one, as it is not given *a priori*, but it is about properties obtained by studying the "form" produced by biological relations.

[13] In the next chapters, after having introduced and reinterpreted the notion of "*closures of constraints*" (Montévil and Mossio 2015), it will certainly be easier to understand how applying the theoretical object of relational biology, namely the relational space.

consider at the same time the constitution of order and the production of disorder shows that organic equilibria of tension have always to face consequences of entropy production. Although these processes are well canalized, as they are subjected to constraints, a symmetry breaking at level of space can radically modify the equilibrium between order and disorder, for example changing the complexity of an organ. It has been noted that even a change of a constraint can *enable* a symmetry breaking.

Anti-entropy, considered as one of measurements showing the stability of organization of a relational space, can be related to tissue complexity. It is very important as it turns out to be another element that shifts the focus of biological research from single cell to "society of cells" (Sonnenschein and Soto 2007). Indeed, cancer research has shown that cancer formation is linked to tissue disorganization (Sonnenschein and Soto 2007, 99 sgg). Therefore its formation has to be studied at level of configuration of relational space and, in particular, of its emerging qualitative characteristics, such as geometry, complexity, etc. (Chollant-Namy and Longo 2022; Montévil 2021). A change of constraints can *enable* a change in cellular reproduction *rhythm* that breaks the equilibrium represented by the mere existence of a tissue. Normally, in the formation of a tumor mass there is an increase in cellular reproduction that, however, builds a structure.[14] Pathology researches (Sonnenschein and Soto 2007, 102 sgg.) have shown that tumors maintain structural bonds with the organ in which they are formed and its metastases maintain similarities with the tissue of first tumor. Hence, the level at which tumors have to be studied is that of relational space, in the sense of relational biology, since cancer and all biological structures are irreducible to the sum of their components. Surely, formation of cancer corresponds to a strong change in values of anti-entropy that reconfigure trajectory of an affected organism. It would be reductive to consider this change only as a disorganization because, not only as the concept of anti-entropy shows, it is not possible to clearly separate order and disorder. Cancer coincides with disorder only from the medical point of view, which is that of an affected organism. From the biological point of view, formation of cancer represents a new attempt of reorganization of a tissue that will also produce entropy in order to maintain itself. It is well known that cellular proliferation and carcinogenic tissue involved are subject to considerable mutations because it is a completely unstable structure since it has no evolutionary history, but it is linked to contingency and to changes in the physical-biological constraints. In fact, "some tumors were successfully transplanted among individuals of the same species, while transplantion into heterologous species, by and large, always failed (Sonnenschein and Soto 2007, pp. 91–98)".

In this sense, besides having organic time*s* and rhythm*s* (plural), there are also "anti-entropie*s*" (plural). Each organism (relational space) produces its own curve of biological entropy, its own irreversibility, starting from configuration of its own relational space and its specific evolution. If death consists in passing from a

[14] This issue will be deepen in the next chapter.

biological system to a closed dissipative physical system, if death is inevitable, the curve that leads to death *can always be redrawn*, depending on how relational space changes, precisely because it represents a possible and an unnecessary path.

Considering studies about anti-entropy associated with the formation of cancer, it is possible to find a clue in direction of conceiving variation as a theoretical principle, since, normally, symmetry breakings modify the possible biological expression: with respect to host organ, *at the level of relational space*, tumor formations can be conceived as hopefulmonsters.

Therefore, considering its generality, anti-entropy is an important tool to deepen stability of biological organization. From the organic point of view, it should be reiterated that production of entropy can correspond to an increase or a decrease in complexity of organ and/or tissue structures. For this reason, as the next chapter shows, anti-entropy will be placed in context of thickness of possibilities. In fact, both development and regression of a tumor are always accompanied by a variation in complexity of tissues. However, all the mathematical data, clinical and technical studies on stability or variation of cancer masses, should be interpreted and the *concept* of thickness allows a general interpretation.

The next chapter shows that this is possible because biological and physical constraints play a fundamental role as they canalize biological processes and allow to study biological objects considering average, respecting the singularity of life.

Bibliography

Badino, Massimiliano. 2010. Ludwig Boltzmann e l'alba della probabilità in fisica. In *Fisica e probabilità*. Milano: Melquíades.

Bailly, Francis, and Giuseppe Longo. 2006. *Mathématiques et sciences de la nature*. Paris: Hermann.

Bailly, Francis, and Giuseppe Longo. 2009. Biological organization and antientropy. *Journal of Biological Systems* 17 (1): 63–96.

Bizzarri, Mariano, and Alessandra Cucina. June 2014. Tumor and the microenvironment: a chance to reframe the paradigm of carcinogenesis? *BioMed Research International* 2014.

Blay, Michel. 1992. *La naissance de la mécanique analytique*. Paris: Presses Universiter de France.

Boltzmann, Ludwig. 2010b. Ulteriori studi sull'equilibrio termodinamico tra le molecole di un gas. In *Fisica e probabilità*. Milano: Melquíades.

Cercignani, Carlo. 1998. *Ludwig Boltzmann. The Man Who Trusted Atoms*. Oxford: Oxford University Press.

Chollant-Namy, Marie, and Giuseppe Longo. 2022. Entropie, neguentropie et anti-entropie : le jeu des tensions pour penser le vivant. https://www.di.ens.fr/users/longo/files/ChollatLongo-entropies.pdf.

Cournot, Antoine-Augustin. 1986. *Exposition de la théorie des chances et des probabilités*. Paris: Hachette.

Gallavotti, Giovanni. 2010. *Statistical Mechanics: A Short Treatise*. Berlin: Springer.

Kant, Immanuel. 1996. *Critica della ragion pura*. Laterza: Roma-Bari.

Kupiec, Jean-Jaques. 2012. *L'ontophylogenèse*. Versailles: Quæ.

Lachièze-Rey, Marc. 2008. *Au-delà de l'espace et du temps*. Paris: Le Pommier.

Laplace, Pierre Simon. 1840. *Essai philosophique sur les probabilité*. Paris: Bachelier.

Leibniz, Gottfried Wilhelm, and Samuel Clarke. 2007. *Exchange of Papers Between Leibniz and Clarke*, ed. Jonathan Bennet. http://www.earlymoderntexts.com/pdfs/leibniz1715_1.pdf.

Longo, Giuseppe. 2016. Comment le future dépend du passé et des événements rares dans le systèmes du vivant. In *La liberté de l'improbable*, ed. Berthoz and Ossola. Collège de France.

Longo, Giuseppe, and Maël Montévil. 2012c. The inert vs. the living state of matter: extended criticality, time geometry, anti-entropy – an overview. *Frontiers in Physiology* 3: 39. ISSN: 1664-042X. https://www.frontiersin.org/article/10.3389/fphys.2012.00039.

Marinucci, Angelo, and Luca Crescenzi. 2021. *Intertestualità. Testo e mondo a partire dalle variazioni*. Pisa: ETS.

Montévil, Maël. 2021. Entropies and the anthropocene crisis. In *AI and Society*. https://montevil.org/assets/pdf/2021-Montevil-Entropies-Anthropocene.pdf.

Montévil, Maël, and Matteo Mossio. 2015. Biological organisation as closure of constraints. *Journal of Theoretical Biology* 372: 179–191.

Newton, Isaac. 2008. *Principi matematici della filosofia naturale*. Milano: Arnoldo Mondadori.

Noble, Denis. 2006. *The Music of Life*. Oxford: Oxford University Press.

Proverbio, Edoardo. 1983. Galileo e il problema della misura del tempo. In *Novità celesti e crisi del sapere*. Fascicolo 2. Supplemento agli Annali dell'istituto e museo di storia della scienza. Firenze.

Rovelli, Carlo. 2008. *Forget Time*. https://arxiv.org/pdf/0903.3832.pdf.

Rovelli, Carlo. 2010. *Quantum Gravity*. Cambridge: Cambridge University Press.

Rovelli, Carlo. 2015. *Is Time's Arrow Perspectival?* https://arxiv.org/pdf/1505.01125v2.pdf.

Rovelli, Carlo. 2017. *A ordem do tempo*. Rio de Janeiro: Objetiva.

Schrödimger, Erwin. 1995. *Che cos'è la vita? La cellula vivente da un punto di vista fisico*. Adelphi: Milano.

Sonnenschein, Carlos, and Ana Soto. 2007. *The Society of Cells*. New York: Taylor & Francis.

Testa, Fabrizio, et al. 2014. Fractal analysis of shape changes in murine osteoblasts cultured under simulated microgravity. *Rendiconti Lincei* 25: 39–47.

Viviani, Vincenzio. 1907. Racconto istorico. In *Opere di Galileo Galilei*, vol. XIX. Firenze: Barbera.

Vulpiani, Angelo, Fabio Cecconi, Andrea Cencini Massimo ancd Puglisi, et al., eds. 2014a. *Large Deviation in Physics*. Berlin: Springer.

Vulpiani, Angelo, Fabio Cecconi, Massimo Cencini, et al., eds. 2014b. *Large Deviations in Physics*. New York: Springer.

Times, Thickness and Relational Space

This chapter represents the core of this second part, because it shows how times and rhythms are treated within relational biology. It was necessary to go through physics and history of science for two main reasons: the first one consists in the lack of theoretical-biological reflections on this subject and, therefore, it was necessary to show a new way to conceive time (Longo and Montévil 2014); the second one refers to the fact that the discussion of concept of time has allowed to show uses and conceptions that are normally obvious, but that, on the contrary, derive from well-defined theoretical choices. Every choice directs inevitably research, opens doors, but closes other possibilities at the same time. In this sense, remaining in traditional biological perspectives, one of the doors that remains irretrievably closed is the possibility to conceive "organism", without surreptitiously reintroducing physical framework into biology (Varela 1979; Moreno and Mossio 2015). On the contrary, this new relational approach allows to construct a solid theory of organism, whose constituent elements can not be reduced to physics, although they are not in contradiction with it. Moreover, through anti-entropy (or better anti-entropies) time can be not necessary considered as the variable *in which* life of organisms takes place because the latter determine its own anti-entropy which, ultimately, is useful tool to consider relational spaces as a whole.

Before dealing directly with themes of this part, it is necessary to summarize some concepts already explained because it is necessary to definitively connect them to the concepts of thickness, relational space and biological times.

In the first part of this text, it has been shown that, starting from the principles of relational biology and applying symmetry breakings to biology, there are *two types of symmetry breakings*, the first one at *cellular* level and the other at level of tissue or any other type of *relational space*. In particular, on the one hand, mitosis produces two different cells, on the other hand, this *does not imply* that symmetry breakings should also occur at level of the relational space, so that, if a cancer cell is implanted in a tissue, it is absorbed or dies (Sonnenschein and Soto 2007). Moreover, it has been shown that specificity of biology is to deal with the *creation* of possibilities,

A. Marinucci, *Theoretical Principles of Relational Biology*,
Human Perspectives in Health Sciences and Technology 6,
https://doi.org/10.1007/978-3-031-39374-7_6

unlike physics dealing with reality. Therefore, it has been stated that *theoretical observable* of relational biology is precisely *relational space*, in *different* forms that it assumes in *reality*: tissue, organ, system, organism, ecosystem etc.[1]

This argument also applies at historical-phylogenetic level because the *irreducible* variety of relational space organizations does not only refer to ontogenetic field, but concerns both existing and past or future phenotypes, considered in one or more species. More generally, as it has been said, relational space *is not constituted* by the relationship between its elements (cells, for example), but it *is* this relationship between its biological elements. Therefore, it can be affirmed that it represents the general form through which biological *organization* can be conceived, at *theoretical* level. Therefore, specific organizations and living organisms are specific types of relational space, at *reality* level.

On closer inspection, a theory that assumes proliferation with variation and motility as principles needs to explain organization.[2] Relational space (in its forms) represents all levels of biological organization since each of them, from the upper level to tissue, is composed of many relational spaces. Being *irreducibly relational*, biological observables can have only *historically* determined[3] structures. In fact, outside relationships, it simply does not exist. For example, this is the case of organism, but, before applying relational space directly to the latter, some clarifications are necessary. As observable, relational space has to be someway measurable, for this reason, considering its definition, measuring a biological relational space means measuring its relations. Concretely, biological times and rhythms (including anti-entropy) can be considered as general measurements of relational space. At this point, it is possible to read again and discuss Noble quote and to interpret it in context of relational biology.

> That, then, is how I came to take my experimental results and a few hand-waving feasibility calculations along to the guardians of the Mercury computer. I must have explained, breathlessly, and in mathematically quite naive terms, what I hoped to achieve. I would fit my data to some non-linear expressions, I thought, then would solve the differential equations for the electrical state of the cardiac cell and then, (hey, presto!), I would see oscillations emerge from the computer's output.
>
> A single question stopped me in my tracks. 'Mr Noble, where is the oscillator in your equations? What is it that you expect to drive the rhythm?'
>
> I was speechless. I had no idea how to reply.
>
> On a sheet of paper, I sketched out the physiological interactions that I thought would work. This served only to confirm my mathematical naivety. There was no oscillator

---

[1] Cell can not be considered the biological observable because it and its functions are such within a tissue. More generally, Even if a cell is considered in itself, it always remains within its context.

[2] A biological theory, based on a physical invariant such as DNA (Monod, Jacob etc.) or organization, has the task of explaining variation, in addition to the necessity to show an invariant element. This is easy in the first case and very difficult, if not impossible, in the second, unless to stated that hurricanes are alive or to go until chemical bonds, but in this last case it is no longer a question of biology.

[3] This topic (Longo 2016) will be deepened when the theoretical place of "organization" is defined in relational biology.

function in my equations. Thirty years later, and after many further rounds of mathematical modelling of cardiac rhythm, I was asked almost exactly the same question by the science correspondent of a national newspaper. By that stage I knew the answer: Silly question! (Of course, I did not put it quite that way to the journalist!)

Yet, it looks like a reasonable question. In a system that oscillates, it seems that there must be some specific component that oscillates, around which the behavior of the entire system is geared; and there must be a mathematical function that describes the way that component oscillates. Indeed, it is an eminently necessary question, if we are talking about some man-made, mechanical systems. But we are not. Instead, we can have a system that operates rhythmically and yet contains no specific 'oscillator' component. There is no need for one. The reason is that the rhythm is an integrative activity that emerges as a result of the interactions of a number of protein (channel) mechanisms.

So there is no need for any oscillatory equation at the molecular level. The rhythm is a systems property. Some biologists have called these properties 'emergent' properties. I prefer 'systems-level' properties, but we are talking about the same kind of phenomenon.

In this quote there are many important elements, but it is enough to focus only on someone of them. Unlike what happens in physics, Noble talks about a system that operates rhythmically without an oscillator; in other words, there is no *subject* of action, described as "rhythmic". Noble states clearly that "the rhythm is a systems property", which means that rhythm (a form of time) is the *result* of an interaction between systems of organism and, therefore, has to described with respect to the shape of relational space which it refers and not with respect to time. Its evolution can consequently be interpreted not with respect to chronological time, but with respect to evolution of relational space that produces it. From the point of view of its properties, these do not change mathematically in function of time, but in function of the elements of relational space: rhythm is a result. In this sense, the geometric description, provided by symmetries, is particularly pertinent.

Noble also points out that rhythm is an "emergent property" or, as he prefers, "systems-level property". This last formulation is actually better as it refers directly to the configuration of the relational space and its levels. Within relational biology, this quotation is also very important because, besides reiterating that heartbeat is a result, it is stated that it is not understandable within a reductionist approach. Now, considering that a biological (relational) space *is not constituted* by the relationship between its elements, but it *is* this relationship between its elements, cardiac rhythm or, more generally, *biological times represent the measurements associated with relational space*, measurements that make it effectively mathematizable. In this context, *thickness is the interpretation of various measurements.*

If relational space *is* the relationships between its elements (or subsystems) and it does not exist without them, it is necessary to explain how it is possible to interpret measurement "time" with respect to the observable of relational biology. In particular, it is possible to state that biological times and rhythms talk about *stability* of relational space (equilibrium of tensions). This formulation it is very pertinent because it shows that equilibrium depends on maintenance of tensions and that it can break at any time. In fact, in biology, possibility of hopefulmonster has to be always kept open, without falling into the Cournot principle (Cournot 1986), since variation is precisely the distinctive character of life. It is important to reiterate that stability (the structure of equilibrium of tensions of an organism or even of a tissue) is not such in an absolute sense, as if there is a principle of a timeless organization. On the contrary, there is stability only with respect to *average* measurements, referred to specific and contingent evolutionary state of a type of organism.[4]

Now, the question is: how is equilibrium maintained, if mitosis always produces symmetry breakings? Why do not symmetry breakings *always* propagate at level of relational space? In order to answer these questions, it is necessary to show how the concepts exposed so far interact and to introduce the concept of organization.

Time, Thickness and Organization

As a measurement resulting from the relations between biological systems and subsystems, time provides informations on stability of relational space and refers to determination of *concrete structure* of a tissue, an organ, an organism etc. *Interpreted as a whole*, these measurements represent the *thickness of possibilities*.[5] In the following pages, it will be a matter to investigate both sides of the question:

1. How stability is related to thickness
2. How stability is structured

As far as the first point, it is a matter of resume what was said about thickness and linking it to time. The second point is much more complicated to deal with, because the real question is: *how is it possible* and, above all, how is *organization* conceivable starting from principles of relational biology (proliferation with variation and motility)? It is easy to understand that many decisive elements are already exposed in order to resolve this issue, however they are not yet sufficient. More properly, biological times and rhythms do *not directly* determine how a relational space is structured or organized, but whether that structure is *stable* or not, without considering how it is. For example, through heartbeat or breathing it is possible to understand if subsystems, producing beat, are stable, compared to past measurements, but its structure can not be understood.[6] Of course, it is

[4] In other words, heartbeat is specific, but, as it will be shown talking about "closure of constraints", contingency of evolution has provided a certain range of heart rates (not binding) for each species. Having a historical-empirical derivation, organization can change.

[5] In the next sections, the relationship between enablement and thickness and how biological measurements characterize the letter will be clarified through examples.

[6] Longo (2016) argues that the importance of past is fundamental in biology. This does not happen in physics, because all the possible states of a system are known *a priori*. In this sense, Longo introduces two extremely important concepts: the "historicized invariance" and the fact that "organisms contextually [interpret] phenotypic traces of the past and re-use them" (Longo 2016, p. 16). The references in this article are philosophy of science and theory of biology, however, the way in which they are developed brings Longo's thought very close to philosophical hermeneutics and intertextuality. It is precisely in this direction that he will deepen these concepts as it is necessary to get out of a logicized epistemology. Therefore, in this book, historicized invariance will be

necessary to deepen its structure or, better, the structures of relational spaces, since biological times are results depending on how they are structured. On closer inspection, each species corresponds to different times and rhythms and, even if within the same species, each living being has his own times.[7] However, in each case biological times, their variation and their stability make possible to clarify the concept of thickness of possibilities, beyond further specifications. Therefore, in order to clarify this issue, it is better to show, first of all, in what sense the measurement of time, stability and thickness are linked and then how to conceive the structure of relational space, i.e. how to conceive biological organization. In this way, the concept of thickness of possibilities can find its definitive place within relational biology.[8]

Time, Stability and Thickness

As mentioned in the first part of this text, the principle of variation allows to conceive that symmetry breakings are possible at all levels in every biological process, even if they are often very improbable at level of relational space (tissues, organs etc.) because of constraints and biological history of organisms. Nevertheless, starting from the principles of relational biology, the possibility of hopefulmoster has always to be kept open. The two sets of possibilities, constituted in every biological process, contains, on the one hand, *historically* known possibilities and, on the other hand, the possibility of hopefulmonster. In such a context, thickness of biological possibilities has been defined as the *measurement* of influence of contingency on the two *sets* of biological possibilities (predictable and unpredictable) such that the latter become *liable* to enter into the domain of reality. The point is that *enablement* shows what is possible, but there is no necessity to pass from possibility to reality. Therefore, it is necessary to interpret the evolution of biological values (times and rhythms) of a living being in order to characterize possibilities quantitatively.[9] Furthermore, it has been shown that biological times derive from the configuration

---

developed starting from Foucault's "historical a priori" (Foucault 2000). The idea of reinterpreting traces will be analyzed starting from a reinterpretation of Agamben's concept of "signature" (Agamben 2008), revisited in theory of intertextuality (Marinucci and Crescenzi 2021).

[7] It is certainly important to investigate diversities and similarities among biological times within the same species, but for the moment it is necessary to proceed further. This aspect will be developed at the same time as the discussion of structure of relational space.

[8] It might seem inadequate to divide the exposition of such an important concept in two, but it was not possible to act otherwise. In fact, in the first part of this book, it was necessary to show that biology deals with constitution of possibilities and with the fact that it is necessary to keep open the possibility of hopefulmonster, since the complete list of possibilities is not given *a priori*. In this second part, the link between biological times, rhythms and thickness would have been incomprehensible without the clarification of concept of biological time.

[9] This issue will be definitively clarified in the section "*Enablement* and thickness".

of relational space and, if they remain stable, it is possible to consider that the structure of relational space *enables* more the production of known phenotypes than hopefulmonster; in other words, thickness of the latter one is low. For the reasons explained in the first part, it is mathematically difficult to apply probability directly to phenotypes because their complete list is not known *a priori*. However, it can be used, *qualitatively*, on the two *sets* of possibilities known and unknown (thickness). Actually, besides being unable to predict whether hopefulmonster will be produced, it is not even possible to determine the type of variation or, even less, the types and/or distribution of known phenotypes (Villoutreix 2015), however it is necessary to keep open possibility of hopefulmonster. For this reason it is necessary to consider thickness of possibilities. Hence, the more historically similar biological times remain, the greater is the thickness of possibilities of historically known phenotypes. Until now, only a general description of thickness has been provided because it is necessary to introduce the concept of organization, as it is configured in relational biology.

The distinction and the relationship between time and thickness is very important in a relational framework, in order to safeguard singularity of living beings. It should never be forgotten that biological object is specific, so biological times and rhythms normally change from phenotype to phenotype, even within the same species, without necessarily involving a change in thickness. The specific structure of each relational space and physical and biological constraints allow that organisms belong to the same species, having *similar* biological times and rhythms *on average*. But this does not imply that the biological object can be reduced to the physical one, because it would be distorted; in fact, rigorously, it would be not talk about "generality" even with regard to species, populations etc. In fact, it would be a big mistake that would prevent the constitution of a theoretical space proper to biology.[10] In fact, from the point of view of variation "similar" is not a degree of "equal", as organicists want, but means "essentially different" (Nietzsche 1999a, 11[166]). This irreducible similarity can be conceived precisely by focusing on biological mean values and from the relational point of view.

Considering any biological rhythm, for each species it is possible to reach values in which it possible to state that a particular organism remains in equilibrium. Of course, like about heart, non-periodic but fractal structure of beat increases its robustness and ability to adapt to different situations. Nevertheless, returning to singular living being, there are cases in which some biological values are outside the "normal" ones, but the structure of relational space does not change. In this case,

[10] As it is easy to guess, introduced concepts lead outside the horizon of physics since, in addition to affirm that it is not always possible to apply probability (mathematical tool) to randomness (philosophical concept), they are characterized by an epistemological indeterminacy: biology is outside determinism that remains at basis of every dynamical approach, in which a purely epistemic uncertainty is conceivable.

biological time of a living being does not respect any average values, but thickness of possibilities remains unchanged.[11]

At this point, an important clarification is necessary: if average values are approximations, thickness is *not* calculated with respect to the statistical average of species, but with respect to the *historical structure of relational space of a singular living being*, whose times complexity *can* be stable. In fact, thickness does not describe directly a *real* singular living being, but the *possibility* that it changes or not. In relational biology, the values of singular living being are irreducible. Therefore, if average values, on the one hand, allow to have information about the *common* elements of a *group* of living beings, on the other hand, *subtract* attention to the *singularity of living beings*, characterizing them as such. In this sense, thickness can only be *specific* because it depends on the stability of a contingent context, in relation to its past history. In this general framework, it is clear the reason why in biology it is necessary to weaken the concept of cause and introduce definitively the one of "enablement" (Longo et al. 2012).[12] This concept ultimately refers to the formation of *new* characters, strengthening the structure of relational space, to the formation of new characters, tending to destroy it, and to maintain the structure of relational space : the distinction between them is given by the concept of thickness. As it will be shown, if the concept of cause belongs to the sphere of reality, the concept of *enablement* belongs to the domain of possibility. Thickness allows to give a value to *enablement* and, therefore, it is a bridge between the domains of possibility and reality one.

The close link between biological rhythms and times, thickness of possibilities and relational space allows to simplify theoretical framework through which biological phenomena and laboratory data are interpreted. Considering the terminology of Noble's quoted text (but this can be considered general) he talks about top-down and bottom-up causation, feedback etc. These are concepts that can be conceived starting from the concept of relational space. From time to time, they specify a particular configuration of a relational space and its functioning, since at each level (tissue, organ, etc.) there are systems which are parts of systems and which are, in turn, composed by systems.

In fact, considering the fundamental principles of relational biology (spontaneous proliferation with variation and motility), symmetry breakings are continually produced in mitosis at cellular level. This situation imposes to pay attention to contingency of each biological process, because it is not necessary that a symmetry breaking is produced also at level of tissue, as happens, on the contrary, in formation of cancer (Sonnenschein and Soto 2007). As mentioned, just tissue (a relational

---

[11] In the last sections of this chapter, thickness will be characterized completely. Now, it is enough to state that different average values change enablement, but it is necessary to focus on history of each living being (measuring times) in order to understand if variation can strengthen organism or destabilizes it. Such a measure is provided by thickness.

[12] In the last sections of this chapter, the possibility to conceive (at least in biology) the concept of "cause" starting from that of thickness will be shown.

space) represents the first biological observable.[13] Considering that tissue is not the unique biological relational space, there are as many observables as there are relational levels in biology, since there are organs, systems and organisms.

Time, Stability and Organization

As said, in-variation of *average* temporal measurements (historically and evolutionary determined) provides the measurement of general *stability* of organisms. Of course, it is worth reiterating that each living being has its own biological times.

So far the relationship between time, stability and *thickness of possibilities* has been deepened, now it is possible to focus on *structure* of relational space, on its *organization*. From this point of view, it is possible to pay attention on the configuration of equilibrium of tensions that constitutes organization, since the latter is always subject to modifications. Hence, in order to conceive organism within relational biology, it is enough to apply what has been said so far. It is more difficult to understand how the organization *is maintained* (and constituted), issue that a theory based on variation has to explain.[14]

In this sense, it is necessary to show the theoretical place of concept of *organization* in relational biology, in order to introduce, *mutatis mutandis*, the *closures of constraints* (Moreno and Mossio 2015)[15] to explain how organization is maintained. In fact, closure of constraints represents an excellent idea to conceive biological organization. This concept is useful to *describe* (not to explain) different forms of organization and to understand what was called "structure of relational space". In other words, in relational biology, the closure of constraints represents the *application* of theoretical concept of relational space.

As mentioned, there are many studies on the concept of organization, starting from Varela up to that most recent by Rosen, Mossio etc. These works share various elements and, in particular, presuppose that organization is an essential element or, indeed, a theoretical principle and then that biology is fundamentally a physics to

[13] About cancer Sonnenschein and Soto (2007) consider tissue as the minimum level for description. Of course, in the theoretical context of relational biology, relational space is the observable, which can be characterized as tissue, organ, organism, etc.

[14] There are numerous and interesting works about organism conceived starting from the principle of organization (Montévil and Mossio 2015; Moreno and Mossio 2015) which are based on reinterpretation of ideas by Varela, Rosen etc. Although this approach is interesting, it fails to construct a theoretical space proper to biology, because, proposing or presupposing a structural invariant (organization), it remains indissolubly bound to physical models, as differently happens for biology by Monod. In fact, although his linguistic approach presents many aspects that do not follow physics (central dogma, exact Boolean algebra, information, etc.), however, his theoretical construction remains physical: for example, Monod himself talks about the need to a "*physical theory of evolution*" (Monod 1989). In both cases, it is a matter of a fundamental invariance.

[15] As for *closure of constraints* see Mossio and Moreno (2010) and Montévil and Mossio (2015).

which are added some *ad hoc* elements to characterize living beings. The problem is that, as in physics, they should find an *invariant* that permeates all living beings, so that they can really consider or implicitly presuppose organization as a biological principle. This is not the place to settle such a question, but this is really a theoretical need. Since *organization is not a principle in relational biology*, it is important to show a new perspective on organization and recover the very interesting idea behind the concept of "closure of constraints".

If organization helps to explain relational structure, from the theoretical point of view, it plays, however, an important role in relational biology. Before discussing closure of constraints, it is necessary to better define how conceiving different forms of organizations, from the relational point of view. In fact, it has been shown that systems are parts of systems that are composed by systems; in this sense, talking about organization or better about *different organizations*, characterizing different phenotypes, relational theory begins to be applied. After all, the burden of proof is on who considers organization as a theoretical principle.[16]

Established cultural and academic habits require to remember again that the *relationships* between the elements of a relational space do not constitute it, but they *are* biological space itself and that biological times are what allows to be able to talk about relations mathematically and biologically and, therefore, about the concept of relational space applied to biology. Moreover, it should be stressed that, systems are strictly referred to relational space that does not exist in itself (as in relativity), but is conceived starting from the *relationship*. Precisely for this reason, it is interesting to conceive it as "equilibrium of tensions" or, like Leibniz, "order of coexistence".[17]

Applying the idea that relational space is biological observable, tissue represents the first observable (Sonnenschein and Soto 2007), while organism is a higher level observable; they correspond and can be described to specific biological times and rhythms.[18]

Relational aspect is very important because, like for the relational space of general relativity (where it is nonsense to talk about empty space, as it derives from the relations between the masses) in biology "organization itself" is not given beyond relations, as it is also a biological-historical empirical or contingent result. Moreover, on closer inspection, removing or modifying all relationships,

[16] In this sense, it is not clear whether closure of constraints forces a particular and general closure or not. One thing is certain: in order to maintain a theoretical coherence, where a principle of organization is admitted or presupposed, it is necessary to propose an invariant structure. If this is impossible, then organization can not be considered as a principle: *mutatis mutandis*, it would be like talking about coding without referring to DNA. From the point of view of relational biology (which does not presuppose any principle of organization) closures of constraints can be reinterpreted as one of the most general and important tools for interpreting and implementing biological theoretical object, the relational space.

[17] In fact, a very interesting thematization about relational time and space can be found in the correspondence between Leibniz and Clarke (2007).

[18] Obviously, this issue can be applied to species, ecosystem etc., because they are relational spaces.

properties, processes, etc. from any relational space (organism, apparatus, tissue, etc.), there are no an "empty" organization, but this latter does not exist. If cancer is developing within an organism, it would completely modify all the relational spaces that constitute it. It distorts completely them, imposing a *new* order of coexistence, a new organization, a new group of properties, a new symmetry.[19]

Considering that relationships are *processes* at every biological level, relational space (and therefore also organism) is structured as an equilibrium of tensions that can be maintained or broken when a break occurs in symmetry at level of relational space. "Organism", "tissue" etc. are ultimately *names*, used at different levels to name relational space. From the philosophical point of view, they are *concepts* that have, from time to time, a different reality and singularity because they are composed of processes having a specific contingent historicity. More generally, every organism, every organization and every relational space are *multiplicity* or, better, a cohesive plurality since its parts are not to be considered as juxtaposed, but as interacting. For this reason, in order to *define* a single organism, it is necessary to *describe*, first of all, *how* it is. In other words, it is very important to always keep in mind that it is a specific object, the result of a possible trajectory, not a geodesic (Bailly and Longo 2006).

Therefore every organism (and every organization) represents a singularity with its own history. Constraints allow to consider and compare *average* measurements of biological times. Although each singularity is irrevocably, it is possible to group living beings in "species", "families" etc., within ontophylogenetic evolution. As already stated by Lamarck and Darwin (Darwin 2011), it is a matter of concepts, useful for talking about groups of living beings, but which do not exist in reality, considering singularity of life. In the same way, there is no organization in itself, since there is *no timeless* invariant, but it is always given within a *specific contingency*. As shown, constraints allow to effectively use these *concepts* if and only if average times are found, including within a minimum and a maximum, as happens for heartbeat, breathing, blood values etc. From the theoretical point of view, all this can be conceived within relational biology as it preserves the singularity of living beings, considering averages as historical and empirical results, but useful for talking about groups of phenotypes. From this perspective, it is again noted that there is no general diagram that can be used as an organizational invariance, it is necessary to admit organization among theoretical principles of biology. Considering the aspect of theoretical coherence, an invariant (especially structural) can vary neither from species to species nor in time. In fact, it is necessary to take into account that biological organization also changes with evolution.

From the *technical* point of view, it could be objected that biological organization reacts by modifying or incorporating typical variations of life, but it remains the central element of biology, as it makes variations possible (Moreno and Mossio 2015). This perspective follows that of Jacob when he states that "changes in the chemical text appear [*surviennent*], not by modification of a previously chosen

[19] Obviously, this new tissue will be very unstable, at least in the beginning of cancer increasing.

sequence, but blindly" (Jacob 1970, p. 289). In fact, variation is always found outside the theoretical structure for organicists and for molecular biology: *from the theoretical point of view*, nothing changes at all, whether the principle is an invariant like DNA or a structural invariant like organization. Obviously, *depending on how concrete biological object is treated*, organicism has the enormous merit of having come out of the prison of code and paying attention to biological structures. For a new biology it is necessary, as said several times, to start from principles that do not follow physics and that do not reduce biology to a particular case of physics.

Where variation is considered as a principle, the problem of theoretical place of organization arises; in fact, it can not be relegated to a subordinate plan, as it represents a central element in any biological approach.

Therefore, in a relational approach, there is the problem about how to deal with organization, because *historicity* plays a fundamental role in creation of multiple forms of organisms. Recusing to propose *ad hoc* hypotheses and to reduce biological organization to the physical one, it is necessary to start from the perspective that allows to construct a *theoretical* biological space. Combining variation and historicity,[20] it is possible to conceive organization biologically.

Organization as Historical-Empirical *A Priori*

If, on the one hand, biological organizations, relational spaces and living beings appear to be *irreducibly singular*, on the other hand, it is important to strongly state that maintenance of different biological organizations (always limited in time) plays a fundamental role also in relational biology. Normally, biologists (with the exception of evolutionists and a few others) are not inclined to consider history as a determining biological element. On the contrary, Longo considers phenotypes as "historicized invariants" and characterizes their evolution through the "reinterpretation of historical traces" (Longo 2016). His way to conceive biology and to express itself shows not only a strong link between biology, hermeneutics and philosophy of history, but he expresses the necessity of a confrontation between science and the so-called "humanities", in which a very long tradition exists. of this concepts. Longo grasps this necessity to get out of biological approaches that relegate historicity to a role of secondary importance,[21] because they are too tied to physics.

[20] The aim is to completely abandon every physical process time or, at least, to subordinate it to biological times and rhythms.

[21] In this sense, it is essential to emphasize the necessity for the philosophy of science to free itself from the chains of an epistemology too tied to logical-analytical approach and to turn to geometry of Poincaré and symmetries. In fact, the latter and Weil (1952) open a physico-mathematical path hitherto little explored, even if the concept of history is completely missing, from the biological point of view.

Considering that in relational biology there are no physical invariants, such as DNA or organization, history *can* finally play a central role. In this sense, in the first part of this text, the concepts of contingency and variation have been explored, now it is possible to propose what derives from them, namely biological organization.

Dealing with an organism or, more generally, with a relational space, it is fundamental to immediately pay attention to its intrinsic historical character, instead of considering it as a physical object (Moreno and Mossio 2015).

Philosophical ideas of Nietzsche and Foucault are extremely useful and, although they can not be discussed in detail, represent a basis to conceive organization within theoretical structure of relational biology.[22] What Longo suggests with the concept of "historical invariance" can be conceived starting from the concept of "historical *a priori*" by Foucault (2000, 170 sgg.)[23] or, more precisely, of *historical-empirical a priori*. Specifically, in biology it is possible to consider historical *a priori* as different types of organization that structure organisms, *from time to time*. In this way, it is possible to safeguard, on the one hand, the existence of different organizations between different species and in evolution of the same species and, on the other hand, the fact that organization is a central element of life. In fact, various forms of organization are the structure and maintenance conditions of organisms, although historical products. Therefore, it is a matter of "*a priori* not of truth [...] but of a history that is given", i.e. no atemporal invariant (no truth) of a physical type is reached through them, as organicists would like, but it is clear that a given species is structured in *a* specific way (facticity).

"In the face of the formal *a priori* whose jurisdiction extends without contingency, it [historical *a priori*] is a purely empirical figure" (Foucault 2000, p. 171). This clarification by Foucault is very important in relational biology because organization can not be conceived as a formal *a priori* because it would mean affirming once again an invariance beyond historical-evolutionary trajectories in which organizations are constituted, especially considering that it is a matter of possible and not of necessary trajectories as in physics.[24]

Starting from the definition of relational space, it is completely contradictory to conceive the existence of an organization beyond its interacting elements: one never goes beyond the historically given organizations, unless one want to build a metaphysics or, worse still, a teleology etc.

Starting from the principles of proliferation with variation and motility and from its consequent contingency, organization is a historical *a priori* in relational biology. The irreducible contingency, typical of relational biology, allows to emphasize the fact that biological historicity is *strictly* empirical, precisely because there is no

[22] Obviously, the concepts of these authors have been significantly modified, in order to adapt them to biology.

[23] Actually, this idea is already in Nietzsche.

[24] Rosen (2005) focuses on metabolic cycles (protein-enzyme-protein), but they are physico-chemical elements, certainly universal, but not rigorously biological, even if present only in organisms: they can be considered as coextensive with biological object. The lack of a properly biological element prevents from constructing imaginary hypotheses beyond biology.

pre-established plan or a general invariance (DNA or organization) able to subsume all types of organization through a common element. Conceiving historicity as a manifestation (even if in different forms) of something that always remains identical to itself,[25] it is however possible to surreptitiously affirm a principle of conservation, thus contradicting the principles of relational biology. From this point of view, every type of organization represents a possible unforeseeable difference and linked to its specific contingency: this *empiricity* expresses the fact that it represents a *possible* trajectory and not a geodesic.

This element is essential to distinguish the theoretical structure of relational biology from all forms of physical or para-biological organicism. In physics, when conditions for the formation of a hurricane are given, this can not fail to form (Longo 2016); on the contrary, in biology, there is never the totality of conditions for which it is something has to arise.[26] At the base of the first type of theoretical structure there is the idea that objects are generic and trajectories are specific, so as in Monod and Jacob, an invariant is sought. At the base of the second and innovative type of theoretical structure there is the fact that object is specific and the trajectories are generic. Summing up with a sentence from Nietzsche: "What is similar is not a degree of equal; but something absolutely different from what is equal" (Nietzsche 1999a, 11[166]).

Although organization is an historical and empirical *a priori*, it still remains an *a priori*, therefore, it shows a form of invariance, albeit a historical one. This type of invariance will be deepened focusing on "closure of constraints"; now, it is possible to affirm that it is always different and is given *by construction*.[27] In other words, the invariance of each organization (for example within species), is given *empirically* and has no value that can be considered so general as to constitute a theoretical principle: as said, "organism" is the *name* given to a certain level of relational space.

If organism and its specific organization are historical and empirical *a priori*, then it is possible to conceive organization as "*idion*" (Aristotele 1955), as "unique property" of a specific organism, which characterizes its singularity, but which *states nothing about the organism itself*, as it is and remains a relational *result*.

From this point of view, organization can not be considered a theoretical principle, *nevertheless* where there is life one meets organization, as it is rightly emphasized by organicists. Of course, it is not (as the latter would like) a physical

---

[25] From the theoretical point of view, it is very different to consider God, DNA or organization as identities subsuming the diversity of forms of life. On the contrary, from the methodological point of view, the structure of thought always remains the same, considering life starting from God, DNA or organization. As shown, relational biology finally tries to conceive life newly. An example of historicity in which historical forms are expressions of a principle is Hegel's philosophy of history (according to certain interpretations) (Hegel 2000, 1997).

[26] It is interesting to emphasize that Nietzsche and Wittgenstein always try to follow this philosophical path.

[27] Here it is easy to understand that the closure of constraints will be reinterpreted and will assume a biological role different from that in the context in which it is born, up to an absolute theoretical incompatibility with organicistic theses.

organization adjusted with *ad hoc* hypotheses to avoid affirming that a hurricane is alive (Varela 1979); it is a different type of organization, because it is conceived in a different theory from the physics one.

If it is possible to agree with what organicists affirm about physical organization, however, in order to conceive a specifically biological organization, it is necessary to resort to the Aristotelian concept of "idion"[28] (unique property), which the Greek philosopher does not refer to biology, obviously. Applying the *idion* to organization, it is possible to affirm that it is an unique property and specific element of organism or of any relational space, but it does not enter into definition of organism. In Aristotle, "unique property" represents a *non-essential* but *coextensive* predicate with the subject to which it refers. For example, "being able to laugh" is a coextensive attribute to "man" because all men are capable of laughing, but it does not enter into the definition of "man". In a relational biology, organization is coextensive with organism or relational space, but does not enter into the definition of organism because the latter avoids any absolute structural invariance, being subjected to history and *empiria*.

Enablement, Thickness and Causality

After showing how organization is conceived in relational biology and after indicating its theoretical *place*, it is important to understand how it is actually configured. Since all kinds of organization are historical-empirical *a priori*, their description can only be contingent. In order to reach a historical-empirical *a priori*, it is enough to put certain elements *shared* by various phenotypes in foreground; however, it is important to stress that this means losing the differences characterizing biological singularity of life.[29] Thus, it becomes possible to consider effectively

[28] Aristotele (1955, 103b1-19): "Let 'same', then, be divided in three as has been said. One proof that arguments are made from and through the things mentioned previously, and are about them, is by means of induction. For if someone were to examine each premiss or problem, then it would be clear that it had arisen either about a definition, or about an unique property, or about genus, or about an accident. Another proof is through deduction. For necessarily, whenever one thing is predicated of another, it either conterpredicates with the subject or it does not. And if it does conterpredicate, then it must be a definition or an unique property (for if it signifies what it is to be something it is a definition, while if it does not it is an unique property - that is what we said an unique property was, something which conterpredicates but does not signify what is to be). But if it does not conterpredicate with the subject, than either it is among the things stated in the definition of the subject or it is not. If it is among the things stated in the definition, then it must be a genus or a differentia, since a definition is composed of a genus and differentia. On the other hand, if it is not among the things stated in the definition, then it is clear that it must be an accident, for an accident was said to be what is neither a definition nor an unique property nor a genus but still belongs to the subject".

[29] According to Lamarck and Darwin, it is necessary to remember that species are concepts that do not exist in reality. Therefore, everything that generally refers to species must be considered, statistically, as averages on singularities, irreducible to each other. This point has already been

species, families etc., but only starting from average values, leaving in background the specificity of each living beings.

In order to explain different biological organizations, it is very interesting to recover the "closure of constraints", as it allows to concretely characterize relational space (theoretical object of relational biology), *not in general*, but in its various forms. In other words, closure of constraints expresses very clearly *how* stability of living beings does not reside in genes or in correctness of a message, but in system, in organization as such (Moreno and Mossio 2015). Nevertheless, complexity of biological structures can not be represented by physical causality (Montévil and Mossio 2015) because in biology there is never the totality of conditions for which something *has to* happen, as in the case of physical self-organization.

Although it is possible to use closure of constraints in relational biology, it is necessary to discuss some properties, because it will be conceived differently. In other words, it is necessary to conceive each closure of constraints neither in a physical framework nor by physical causality.

A first general and theoretical difference imposes immediately itself. In organicist approach, organization is considered (or implicitly presupposed) as a biological principle and, therefore, variation acts, *from the outside* on the closure of constraints reinforcing it or dissolving it. In this case, as for Monod, variation lies outside the theoretical framework. On the contrary, starting from the principles of relational biology, proliferation with variation and motility, it is possible to actually conceive organization as a set of structured constraints that *limit and canalize* variation, without being able to block it altogether. Just the fact of considering variation as a theoretical principle and organization as an historical-empirical *a priori* allows to distinguish the *use* of closure of constraints in relational biology from the organicistic (physical) perspective. This is a strictly theoretical choice, which focuses on the way to conceive theoretical biology.[30]

Although closure of constraints allows to describe *quantitatively* organization, however, there are other elements that differentiate its use in relational biology. As said, there are always only closure*s* (plural) of constraints, because it is not necessary to assume the existence of any structural invariant (organization itself) subsuming all living beings: their organizations are historical-empirical *a priori*.

In particular, organicists (like Varela) treat biological organization along the lines of classical physics by adding *ad hoc* elements, trying to reach something strictly biological. Proof of this is the pervasive and structural use of causality to describe organization (classical physics like), whose description of reality is and has to be mathematically deterministic. For this reason, the work of Montévil and Mossio on closure of constraints (Montévil and Mossio 2015; Moreno and Mossio 2015) is extremely important as it appears to distance itself more and more from physics and

---

clarified regarding the average values of biological rhythms and times; the same argumentation applies to the descriptions of biological organizations.

[30] In this sense, the opposition between "somatic mutation theory" and "organization field theory" is very similar (Sonnenschein and Soto 2007).

to approach variation and contingency, of which relational biology alone constitutes the theoretical basis.[31]

As mentioned, the element that varelian organicists do not consider adequately is that, according to a physical description, a hurricane *must necessarily* form, when its conditions are given: the set of causes that produce it are *necessary and sufficient*. On the contrary, in biology, since its object is specific and due to intrinsic theoretical indeterminacy, the totality of conditions for which something *should* happen is never given, as hopefulmonster is always possible. From the point of view of dynamic systems, when one studies a hurricane (or a finite number of types), all the emergent properties that characterize these atmospheric phenomena are known, although the latter are not reducible to singular causes, but refer to them *as they form a closure of constraints* (Moreno and Mossio 2015). In other words, physical object is generic and the indeterminacy is epistemic. However, in biology, emerging properties are *not all* describable starting from presumed "causes" or their interaction, since hopefulmonster remains indeterminable,[32] as for its properties: biological object is specific and indeterminacy is epistemological.

From the point of view of relational biology, the use of a physical causality is theoretically very weak, precisely because it fails to account satisfactorily for the possibilities of biological processes (Longo et al. 2012). Nevertheless, closure of constraints remains an essential tool *to describe* organization.

Cause and Enablement

At this point it is necessary to dwell on *causal* closure of constraints and on critical use of physical causality in biology. Of course, causality will not be discarded, but considered differently. In this regard, it is important to clarify briefly how it interact with the concept of "enablement" (Longo et al. 2012) and with that of "thickness", precisely in order to safeguard possibility of hopefulmonster and because, in relational biology, "closure of constraints" is interpreted as *limits* it

[31] It is no coincidence that there is greater attention to the principle of variation and to contingency (Mossio et al. 2016): "A given biological organization is determined by an accumulation of changes of symmetries both on the evolutionary and the ontogenetic times. These changes correspond to changes in the manner in which functions are performed, or even to the appearance or loss of functions The crucial consequence of this view is that, because of their permanent symmetry changes, biological objects should not be considered as generic objects. Organisms are not well defined as invariant under transformations". At least from the point of view of relational biology, considering this series of texts, it is incoherent to combine the principle of variation and that of organization.

[32] It has been said that this indeterminacy is intrinsic in the sense of quantum mechanics, therefore, it is not an epistemic limit, which can be overcome if all the elements involved were known, as conjectured by Einstein, but it is a matter of an epistemological uncertainty. From this point of view, it is important to remember that it is not theoretically enough to leave radical variation outside the theoretical field of biology, as organicists and biologists, following ideas of Monod, do.

and *canalizes* variation, without suppressing it. Biological variation does not act *externally*,[33] as if the state of default of life is *quiescence*.[34]

In general, it is possible to affirm that causality represents, kantianally, a possible way to construct a correlation between two successive events. If B follows A, a causal relationship means affirming that $A \Rightarrow B$. In classical physics, once differential equations have been written, it is possible to identify the responsible of a given effect, establishing the cause-effect relationship between A and B. In biology, on the contrary, cause can be considered as "differential":

> A classical mistake is to say: this mutation causes an idiot child (a famous genetic disorder, phenylketonuria), thus... the gene affected by the mutation is the gene of intelligence, or: here is the gene that causes/determines the intelligence. In logical terms, it consists in deducing from «notA implies notB», that «A implies B»: an amazing mistake. All that we know is a causal correlation of differences (Longo et al. 2012, p. 14).

Classical physics establishes a direct relationship between equations and phenomenon described, so that everything calculated is *liable* to enter into reality, therefore, all the possibilities of a physical system are "flattened" on reality, they are, so to speak, possible or potential reality. In a more rigorous form, it possible to state that the general solution of an equation encompasses all the possible *real* dynamics, then boundary conditions of phenomenon determine its particular solution. Thus, differential equations are mathematical expressions of all physical causes involved in a system.

The acritical application of physical causality to biology reproduces the same situation. However, from the theoretical point of view, these two disciplines refer to different domains or, at least, relational biology does not intend to reduce biological possibility to the physical one, always given *a priori* and always *liable* to enter into actual *reality*.

In order to construct correlations in the domain of reality, scientists normally pretend to apply acritically an instrument (causality) to a science (biology) which deals with possibility and its constitution. From the physical point of view, closure of constraints is structured by causal structure and, therefore, builds relationships at level of *reality*, eliminating completely the question of enablement and variation from *theoretical* biology.[35] In order to consider enablement and variation from the

---

[33] Obviously, from the practical point of view, it is possible to consider an external causality, but it has to be considered relationally. From the theoretical point of view, it is necessary to conceive a *biological* causality.

[34] The authors of *The Society of Cells* insist a lot on the necessity to abandon *quiescence* and adopt *proliferation* as default state of life. From the philosophical point of view, assuming a physical invariant, it is impossible to escape quiescence as a starting point for any type of biological investigation (where, clearly, theoretical coherence is considered an important value). Where the principles of proliferation with variation and motility are assumed, it is completely contradictory to assume organization as a principle, since quiescence would be reintroduced.

[35] In this sense, it is not enough to state that in biology a symmetry breaking "survient" to DNA (Monod 1989; Jacob 1970) or to biological organization (Moreno and Mossio 2015). In order to construct a new biology, variation can not more exclude to its theoretical framework.

theoretical approach, biology has to have an own domain of possibility not reducible to the domain of reality, as in classical physics.[36]

Enablement and Thickness

As shown, if physical causality acts at level of reality, biological enablement acts at level of *possibility* and its *constitution*. The concept of enablement expresses the fact that a set of elements *makes* something *possible*, but does not causes it. Closures of constraints make possible nothing but *describe* different real biological organizations and show how they act on variation, *limiting* it. Closure of constraints is, ultimately, a quantitative tool characterizing various *real* historical-empirical *a priori*, according to their different structures.[37]

If the set of historical-empirical *a priori* and the contingency of environment *enable* certain possibilities, then enablement (unlike causality) *does not* act at level of *reality*, but at that of biological *possibility*, i.e. it constitutes *contingent* possibility. In other words, enablement does not allow establishing any causal and/or necessary relationship between possibilities and what is measured: it is a *qualitative* tool.

Established that enablement acts on possibility, it is necessary to introduce the thickness of possibilities both to understand how it is possible to pass from *contingent possibility* to *potential reality* and how conceiving causality biologically. This second aspect is very important. In fact, if biology is reduced to physics and physical causality and necessity are used in biology, there is necessity to introduce neither enablement nor thickness; besides, both orthodox biology and organicism leave possibility of variation outside the theoretical framework,[38] referring only to potential and actual reality. In order to explain how conceiving biological causality starting from thickness, it is necessary to clarify the latter preliminary.

Enablement constitutes qualitatively the concrete or contingent possibilities of a biological process. In general, they appear as set of predictable phenotypes (P_p) and set of unpredictable phenotypes (P_h).[39] Even West-Eberhardt argues, selection acts on a biological "random", always canalized by contingency (West-Eberhard 2005,

[36] Quantum mechanics shows this necessity (Heisenberg 1971), because there are possibilities not liable to enter in the realm of reality (i.e. quantum superposition). In biology, theoretical situation is more complex because the complete list of possibilities is not given *a priori*, differently from classical and quantum physics.

[37] It is not necessary to repeat the arguments of aforementioned organicists to reiterate the importance of closure, even if not strictly causal.

[38] As the title of Monod's famous book summarizes, there are two distinct domains in biology: chance and necessity.

[39] It is necessary to express in this way because possibilities are always contingent with respect to a specific biological description and not with respect to an abstract and imaginary "possibility in itself", not better specifiable.

2003). Enablement constitutes and directs possibility starting from an irreducible contingency.

It should be pointed out that, when it is said that certain possibilities are enabled, it is clear that it is a matter of *an interpretation* of a biological system, which can not exclude other interpretations. This means that, although it is possible to direct hopefulmonster prediction by enablement, it is not possible to exclude other forms. In other words, it is about the level of description and the level of possibility and its constitution and not the level of reality. In this sense, no kind of cause can determine a phenotype. Due to this intrinsic and irreducible indeterminacy, thickness is applied to the *sets* of possible known phenotypes (P_p) and unknown (P_h) and not to the *possible* phenotypes.[40] Hence, as mentioned in the first part, thickness provides the *measurement* of influence of contingency on the sets of biological possibilities (predictable and unpredictable), such that the latter become *liable* to enter into domain of reality. Expressed in the *relation* between P_p and P_h, this measurement is given by biological times, in so far they can be interpreted with respect to the stability of a relational space.[41]

In general, if P_p indicates thickness of *set* of predictable phenotypes and P_h that of *set* of non-predictable phenotypes,[42] in relational biology applies:

- $P_h \gneqq 0$ and $P_p \gneqq 0$
- If $P_p \gg P_h$, possibility of no variations at space level (organization) is more robust than variation, without enablement's changes.
- If enablement changes, but $P_p \gg P_h$, then robustness of system increases.
- Classical causality and necessity (that it conveys) can be conceived *as if* $P_h = 0$, but since in biology $P_h \gneqq 0$, it is possible to talk about causality, in a non-rigorous form, only when $P_h \to 0$

Causality has been interpreted in this way in order to emphasize the irreducible difference between classical physics and biology. In fact, it is different to state that the hurricane *must* form, given certain conditions (causes), and state that something *can* be formed, given precise conditions. In the first case, object is considered epistemologically in the domain of reality; in the second one, in that of possibility. In this sense, causality belongs to the domain of reality, enablement and thickness to that of contingent possibility; in particular, the latter is involved in transition from contingent possibility to potential reality.

---

[40] Obviously, one of the two sets contains historically known phenotypes, but they are considered as possibilities and not as realities. Now, one of the most important elements of irreconcilability between physics and biology is proposed: if the domain of the former is reality, the domain of the latter is possibilities.

[41] As it has already been shown, this element is extremely important because it is possible to consider two living beings from the same species with the same enablement, but with different thicknesses.

[42] Here, "h" means *hopefulmonster*. In this case, it refers not only to phenotypes, but also to any modification of any relational space. For example, P_h also indicates the formation of a cancer or the onset of a disease.

What has been said so far allows to understand the status of enablement and thickness, but an example may help. A doctor[43] who diagnoses a viral or bacterial disease rightly states that an external element "causes" disease because this would not have arisen without virus or bacteria. Nevertheless, he has to study the conditions for which that disease could have developed, as there is no need for which, in the presence of viruses or bacteria, a disease *must* manifest itself. From the deterministic point of view of classical physics, if presence of external agents are a necessary and sufficient condition for development of disease, then everyone should always be sick or, at least, everyone should get sick starting from the same initial conditions, which is biologically completely absurd, but physically understandable. In this case, considering unchanged enablement, what changes is thickness, $P_h \gg P_p$, until a "cause" appears as such.[44]

"Causal" closure, creating and maintaining a hurricane, is different from those characterizing various kinds of biological organizations. As for the rhythm of heart, as shown by Noble's work summarized in part in the quotation proposed, not only oscillator ("cause") does not exist, but there are never the same conditions for different living beings. Furthermore, a qualitative change in causal regime, that produces and maintains (and does not make possible) a hurricane, determines its breakdown. In biology, a qualitative change of a closure of constraints can increase its robustness, depending on thickness measurement.[45] As seen, one of the fundamental differences between physics and biology, between physical and biological autopoiesis, is that in the latter *closures of constraints* describe structures and *limit and canalize* variation;[46] on the contrary, in physics, closure of constraints creates a structure, does not limit anything and expresses a necessity linked to a specific causal regime.[47]

[43] This example is discussed in Longo (2018).

[44] An extreme example certainly helps to understand the differences between thickness and causality. If a very powerful poison is injected into an organism, it dies undoubtedly. This " necessity " can be explained by the fact that $P_h \gg P_p$ and $P_p \to 0$. As mentioned, $P_p \gneqq 0$ means that the possibility of surviving the poison is always present even if minimal. It is a situation similar to that occurring in statistical mechanics when it is possible to state that it is not impossible, but only extremely improbable a spontaneous transition from disorder to order. On the contrary, biological approaches, assuming a fundamental invariant as conservation principle, can not explain cancer without introducing a physical "cause". As it will be shown, through the analysis of a scheme taken from *The society of cells*, an endocrine disruptor disrupts the control of reproduction, but does not produce cancer, *enables* it, because relational biology is deals with the constitution of possibilities through the study of stability of relational space.

[45] From the theoretical point of view, it is not enough to point out that modifications can increase robustness in biological systems, it is necessary to derive it theoretically, as happens in relational biology. This element is crucial because if Newton takes up Kepler's laws in his *Principia*, in Kepler it is a matter of observational laws, on the contrary, Newton derives them from theoretical body of *Principia*, exposed especially in the first book.

[46] It is well known that increase or decrease of gravity affects leavening process.

[47] this element is partially present in organicism approach, but it is not clear how to conceive the fact that, even when speaking of causes, there is no necessity in biology. There is no a theoretical space that allowing it.

The use of closure of constraints in relational biology requires a further study about causality, starting from what has been said about enablement and thickness. In general, in relational biology the closures of constraints are *not* governed by a physical causality, but by a closure for which $P_p \gg P_h$ with $P_h \to 0$, but always with $P_h \gneqq 0$. The latter is conceivable exclusively starting from the principles of relational biology; again, closure acts on variation limiting it, it is not (only) about a variation acting from the outside on closure of constraints. Some external element can be free biological variation.

Up to now, enablement and thickness have been discussed from the point of view of theoretical structure of relational biology, as it presents remarkable differences with respect to physics. Nevertheless, it may seem difficult to understand its actual application, therefore, it is necessary to discuss it from another point of view, for example the medical one.[48]

In *The society of cells* a schematic representation of cancer formation process (Fig. 1) is proposed according to *Tissue organization field theory* (Sonnenschein and Soto 2007, p. 105). It is very interesting because it can be interpreted within relational biology, since the principles of proliferation with variation and motility are also adopted in this text.[49]

It should be remembered that in the case of SMT (somatic mutation theory) there would be a scheme like this:

$$Carcinogen \Rightarrow Hyperplasia \Rightarrow Dysplasia \Rightarrow Carcinoma \\ \Rightarrow Clinical\ cancer.$$

Comparing to Fig. 1, the arrows returning back and showing the possibility of disease regression are missing. Consequently, the trajectory of SMT is necessary and causally determined. On the contrary, in relational biology trajectories are possible, therefore, besides the direction of SMT, it is necessary to add the *possibility* of cancer *regression*. In other words, as shown by the arrows in the scheme proposed by Soto and Sonnenschein, enabled possibilities are "health" and "disease" (stability and instability of relational space). In this sense, biological object is characterized qualitatively and contingently: with respect to the "normal" structure of a given tissue, a certain enablement is created (level of possibility). Now, the point is that at every level (hyperplasia, dysplasia etc.) enablement remains the same, because the arrows go in the direction of disease and in the opposite one, however, *something changes*: since there is no necessity, it comes into play the *thickness* of possibilities.

[48] Naturally, from this point of view, it is possible to replace the theoretical terms "stability" and "instability" of relational space with "health" and "disease". Obviously, from the point of view of cancer formation, this perspective would be different.

[49] Considering the theoretical character of this text, it is not necessary to re-propose experimental data of the research group of Soto and Sonnenschein, duly referred to in the bibliographies of *The society of cells*.

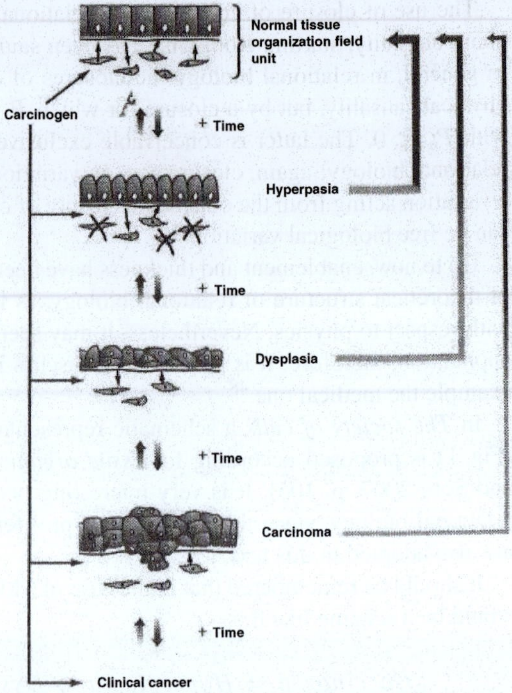

Fig. 1 Schematics representation of the process of experimental carcinogenesis according to the *tissue organization field theory*

Considering as *distinct and separated*, changing values of disease are *quantitative* informations, but interpreted as *distinct and inseparable*,[50] they are able to express if tissue stability increases (robustness) or decreases, i.e. if $P_p \gg P_h$ or $P_h \gg P_p$. Enablement remains unchanged between hyperplasia, dysplasia and carcinoma, but something can change (for example, the patient gets worse); in this sense, the *relation* between P_p and P_h changes: the relation between what is *enabled* is quantitatively characterized. There is a switching from $P_p \gg P_h$ a $P_h \gg P_p$, From "normal" tissue to carcinoma or to histologic examination. The results of clinical examinations that detect hyperplasia, dysplasia, carcinoma etc. represent quantities that can be interpreted as the relationship between the enabled *possibilities* (P_p and P_h), because otherwise there would be no arrows, for example, from carcinoma to normal situation or to dysplasia (Fig. 1). Therefore, *thickness* allows to characterize not exactly the evolution of the two *sets* of possibilities P_p and P_h (normal tissue and disease), but their *relationship*, because there is no kind of process time[51] nor a necessity linked to it. In fact, the same applies from

[50] Regarding biological elements, conceived as "distinct and separated" or as "distinct and inseparable", please refer to what has been said in the first part and to what Noble affirms in the quoted passage on the lack of an oscillator for heart rhythm. For a philosophical explanation: Marinucci and Crescenzi (2021).

[51] About time of processes and time of life see Longo (2016), Montévil (2022).

cancer to dysplasia, it can not be said that it is impossible to return to carcinoma, therefore, the possibility of disease is not erased, but the *relationship* between *possibilities* is modified (from $P_h \gg P_p$ to $P_h > P_p$). As it is easy to guess, it is at *level of possibility*: enablement has allowed to qualitatively constitute the two sets of possibilities P_p and P_h, while thickness has allowed to characterize them quantitatively. On the contrary, the "*measurement*" of cancer belongs to the realm of reality.

In general, it can state:

- Historical-empirical *a priori* (average evolutionary values specific to a tissue or relational space), evolutionary contingency.
- Values outside the average of singular living being (Hyperplasia).
- Enablement: *qualitative* characterization of possibilities, starting from evolutionary contingency and from individual values: constitution of the two sets of the possible (P_p e P_h).
- Thickness: *quantitative* characterization of possibilities or development of singular contingency: the development of tissue is considered not in reference to average (enablement), but to the singularity of living being, to its development, for which P_h e P_p can be related.[52]
- Measurement: clinical detection of cancer (actual reality).

At this point, it is useful resuming and complete the scheme proposed in the second chapter of first part concerning the domain of biology:[53]

Contingent possibility {
- Historical-empirical *a priori* (Evolution)
  - *Enablement* (costitution of possibilities)
- Singularity of living being
  - Thickness (Relation and development of possibilities)

Contingent reality {
- Potential or possible reality
  - Mesurement
- Actual reality

It is interesting to stress what was said using the article *Carcinogenesis explained within the context of a theory of organism* (Sonnenschein and Soto 2016). Always dealing with the formation of cancer from the point of view of tissue, Soto and Sonnenschein affirm that cancer is enabled by an alteration of tissue organization and by an excessive accumulation of cells, practically, it is a "development gone awry" (Sonnenschein and Soto 2016, p. 3). In other words, any relational space,

[52] This does not prevent new diseases from occurring, because biological processes are not characterized by necessity, but by contingency.

[53] It is very important to stress that "thickness" and "measurement" are not part of biological possibility or reality, but allow the transition from one level to another.

for example an organ, is a historical-empirical *a priori* with its own structure described as closure of constraints. Its development and its morphogenesis are related to how its constraints are able to *limit* proliferation with variation and motility, characterizing the default-state of life.

> We posit that the loosening of any of the constraints described in the morphogenesis model may lead to abnormal tissue organization, and if persistent, to carcinogenesis (Sonnenschein and Soto 2016, p. 4).

In this case, from the conceptual point of view, there are two moments that it is important to keep distinct: 1. the "loosening" of some *constraints* and 2. the persistence ("if persistent"). The first one enables a certain type of cancer while the second one requires to follow tissue development in order to consider if it goes *towards* cancer or not.[54] The loosening and the consequent enablement indicate the "direction" in which that tissue has to be considered, namely health-disease or, better, stability-instability of its relational space. Since in biology there is no necessity or physical causality,[55] *thickness* is between possibility and reality. A new element or a new possible (enabled) phenotype represents a *qualitative* element, whose development has to be observed, in the *direction* of its enablement.

If enablement provides directions of tissue development, thickness provides its *verse*, but not of disease, of tissue development because enablement belongs to the domain of possibility, while the appearance of disease belongs to the domain of reality. Since there is no need for either physical causality or procedural time between these two domains, it is necessary to follow tissue development because cancer can stop and/or regress. Following the development of tissue, enablement is presupposed as it shows biological possibilities, while thickness is the *measurement* of influence of contingency on the two *sets* of biological possibilities (predictable and unpredictable) such that these become *liable* to enter into domain of reality. As mentioned, this measurement is necessary because cancer *can*[56] regress (Sonnenschein and Soto 2016, 2007).

> ...it is worth recalling instances where, on the one hand, normal tissues transplanted into the "wrong" locations resulted in neoplasia, while, on the other, genuine cancer tissues and their cells became normalized after being placed in the midst of normal tissues (normal niches) (Sonnenschein and Soto 2016, p. 5).

In the case of cancer regression, its possibility does not disappear, but it is always enabled, however there is a change in biological *values* so that the relation $P_h \gg P_p$ becomes $P_p \gg P_h$.

[54] It is important to remember that there is no necessity and, therefore, it is impossible to use necessity and/or physical causality uncritically and dogmatically.

[55] It should be remembered that physical causality can be introduced in biology as a borderline case for which there is always $P_h \gneqq 0$ and $P_p \gneqq 0$, with $P_p \to 0$.

[56] The use of this verb ("to can") means that there is no physical necessity or causality. As mentioned, in biology there are at most the extreme situations in which $P_p \gg P_h$ with $P_h \to 0$ or in which $P_h \gg P_p$ with $P_p \to 0$.

On the contrary, "once a cancer cell always a cancer cell" shows an approach for which there are no longer any distinctions between possibility and reality and shows that everything is considered in the domain of reality. If "once a cancer cell always a cancer cell", then it is not necessary to follow the path of disease because, in this case, enablement and reality of disease coincide. Similarly, it would no longer make sense to talk about the relationship between P_h and P_p because biological development would become a physical development with a procedural time, but physics is not biology and the latter has its own concepts that make no sense in physics. However, since the SMT approach is no longer sustainable, it is necessary to formulate a new theoretical framework with new concepts such as, precisely, enablement and thickness are able of accounting for the constitution of possibilities and its development that of relational biology, which presents itself as a radical theory of ontophylogenetic contingency.

Thickness and Causality

If biological causality has been conceived starting from thickness, this does not mean that it is necessary to eliminate it. In relational biology, it is configured differently from physics, even if there is no contradiction between them. As said, this difference derives from the irreducibility of biological object to the physical one and from its theoretical structure: if physics is based on something that is conserved, biology on something that is *not* conserved, on symmetry breaking. Hence, biology deals with possibility and its constitution, while the domain of physics is reality.

The example of doctor shows clearly that in biology it is not rigorous to use causality in a physical sense. In fact, a virus can have an effect, it can assume the role of "*cause* of disease", only if it is possible to consider thickness $P_h \gg P_p$ (and $P_p \rightarrow 0$). Naturally, disease is enabled insofar as there is a change in the relational space (organ). It is possible to follow its development through the relationship between P_p and P_h, up to measurement, up to compromising or re-establishment of stability of an organ.

At this point, two elements are important, the first one is that, unlike hurricanes, biological conditions for disease can be the most disparate. The second one, closely related to the first, is that the appearance of disease occurs when $P_h > P_p$ or, even, $P_h \gg P_p$. It is, rightly, the contingent variation of stability of a relational space, such as to *allow* the virus to become the "cause" of disease. Since $P_p \gneqq 0$, under the same conditions, it is not necessary to get sick or have always the same health problems; under the same conditions it is even possible that a new disease arises. The central element is the stability of an organ, of a tissue, linked to the patient's past contingent history. In other words, strictly speaking, it is not so important if

$P_p \to 0$, but that $P_h > P_p$. In this case, if $P_h > P_p$, then there is a greater thickness that disease can appear.[57] The interesting element is that, starting from thickness and, above all, from $P_h > P_p$, a virus, a bacterium etc., *can* play the role of "cause" of a disease or, more generally, *can* destabilize a relational space. Considering that in all organism viruses and bacteria are always present, rather than the "cause", the central element is precisely thicknessTherefore, causality in biology is *not* linked to necessary and sufficient conditions, but to thickness, excluding the possibility of necessary and sufficient conditions.[58] It is clear that closure of constraints represents a structure for which possibilities are strongly canalized, but this does not eliminate the consequence of theoretical principles of relational biology, for which $P_h \gneqq 0$ is always valid. The fact that the domain of biology is possibility and its constitution imposes itself again.

What has been said allows to broaden the use of relational biology. This becomes extremely clear, for example, in all those diseases in which one does not deal with a "cause", such as viruses, bacteria or other elements external to organism. The most interesting example is cancer.

At least since the 70s, based on a theoretical biology allowing *no* theoretical possibilities different from the physical ones (Monod 1989; Jacob 1970; Varela 1979; Deutsch 2012), the "cause" of cancer has been sought, as a retro-virus, a bacterium. It was also attempt to cure cancer reprogramming the code of diseased cells as in *Somatic mutation theory* (Nowell 1976; Weinberg 2014). Unfortunately, these failures depend *causally* from the theoretical perspective limited by physics to the domain of reality.

The book *The society of cells* was already mentioned, but it is important to underline the importance of the principles of proliferation with variation and motility.[59] In general, it can be said that cancer has neither "external causes" to tissue or organism nor "internal causes", in the sense of physical causality. An endocrine disruptor does not "cause" cancer (domain of reality) in a certain organ, but modifies enablement (domain of possibility). In this sense, necessary and sufficient conditions or causes for appearance of cancer are never present. If closure of constraints were really ruled by a causal and physical-like regime, then cancer should never occur. In the theoretical space of closure of constraints, without

[57] Obviously, it makes no sense to talk about "more possible" or "more enabled" (qualitative elements), but it makes sense to talk about thickness (quantitative element), because the latter is measurable. Furthermore, if enablement refers to the level of constitution of possibilities, thickness refers to the transition from contingent possibility to potential reality, focusing on singularity of each biological process.

[58] In this case too, it is possible to talk about necessary and sufficient conditions (not rigorously) only for $P_p \to 0$. It is important to stress that it is precisely the general theoretical context of relational biology that opens spaces to conceive a new biology, that were previously relegated to simple practical but extra-theoretical elements, such as variation.

[59] The theoretical core of the differences between biology and physics is shown in Bailly and Longo (2006).

enablement and thickness, variation can only be explained by an external agent, which does not always exist in the case of cancer.[60]

For example, it is possible that a heavy smoker develop no type of cancer, while who does not smoke and who lives in a healthy environment manifests it. These kinds of examples highlight the question of possibility and applicability of notion of physical causality. In spite of the fact that for the non-smoker is worth $P_p \gg P_h$, the disease *can* be generated ($P_h \gneqq 0$). On the contrary, in the case of smoker, $P_h \gg P_p$, but despite this situation does not require anything, there is a quantitative measurement available which gives qualitative information ("towards" it has been said) about the difference between the smoker and the non-smoker.

From the point of view of relational biology, cancer can be conceived starting from the fact that historical contingency enables a specific organization, for example that of lungs. This means that there is always $P_h \gneqq 0$, with $P_p \gg P_h$ and *not* that, since there is a normal lungs structure, $P_h = 0$, because the closure of constraints limits variation (default-state), it is not the latter that messes up the first. The sensitive increase of P_h ($P_h \to P_p$) does not produce *necessarily* a symmetry breaking at tissue level, but testifies that something is changing and that the chances of a variation are increasing ($P_h \gg P_p$). However, this depends not only on *enabled possibilities*, but also on *real* history of singularity of life, object of study, as the example of smoker shows.[61] For an organism of a smoker, it is also possible to find an organization for which, despite cancer is enabled, his thickness remains $P_p > P_h$. The central point is that, while causality belongs to the domain of reality, thickness belongs to that of possibility and manages transition from possibility to potential reality.

What has been said so far shows the role of "causality" in the theoretical framework of relational biology, in addition to show the necessity for enablement and thickness. It can be considered as the element that *triggers* a symmetry breaking, which enables a new trait that does not immediately pass to reality just because it is possible, but it is necessary a modification of thickness, because this is the only way to conceive the "towards" of enablement.

If causality can be conceived as the triggering element, thickness shows (as seen with respect to evolution Carcinogen $\Rightarrow$ Clinical cancer) how a certain biological relational space evolves because what is enabled may or may not pass into domain of reality. In the case of *real appearance* of a disease, this symmetry breaking does not give rise to an "*effect*", but to a *reconfiguration of relational space*, i.e. to structuring

---

[60] It could still be objected that asbestos, as a disruptor of the tissue structure, can be considered the cause affecting cellular connections that canalizes reproduction with variation. Actually, disruptors do not act on level of reality, but on that of possibility; for this reason, enablement and thickness are necessary to explain their action. Disruptors do not cause cancer, they enable cancer. In fact, it may not manifest itself. Furthermore, as the diagram from *The Society of Cells* shows, the possibility of cancer regression is always enabled. In this sense, it is necessary to follow the development of stability of relational space through thickness.

[61] It is important to stress that in relational biology it is possible to talk about necessity when $P_p \to 0$ with $P_h \gneqq 0$ or $P_h \to 0$ with $P_p \gneqq 0$. Obviously, it is not about physical necessity.

of new properties (symmetry), just as occurs in cancer. In turn, this reconfiguration will have a specific enablement, which can be connected to a thickness etc. The massive variation in cancer tissue is due to the lack of a structure of constraints, limiting biological variation.

Bibliography

Agamben, Giorgio. 2008. *Signatura rerum*. Torino: Bollati Boringhieri.

Aristotele. 1955. Topici. In *Organon*. Torino: Einaudi.

Bailly, Francis, and Giuseppe Longo. 2006. *Mathématiques et sciences de la nature*. Paris: Hermann.

Cournot, Antoine-Augustin. 1986. *Exposition de la théorie des chances et des probabilités*. Paris: Hachette.

Darwin, Charles. 2011. *L'origine delle specie*. Torino: Bollati Boringhieri.

Deutsch, Jean. 2012. *Le gène. Un concept en évolution*. Paris: Seuil.

Foucault, Michel. 2000. *L'archeologia del sapere*. Milano: Bompiani.

Hegel, Georg W. F. 1997. *Lezioni di filosofia della storia*, vol. 1. Firenze: La Nuova Italia.

Hegel, Georg W. F. 2000. *Enciclepedia delle scienze filosofiche*. Roma-Bari: Laterza.

Heisenberg, Werner. 1971. *Physique et philosophie: la science moderne en révolution*. Paris: A. Michel.

Jacob, François. 1970. *La logique du vivant*. Paris: Gallimard.

Leibniz, Gottfried Wilhelm, and Samuel Clarke. 2007. *Exchange of Papers Between Leibniz and Clarke*, ed. Jonathan Bennet. http://www.earlymoderntexts.com/pdfs/leibniz1715_1.pdf.

Longo, Giuseppe. 2016. Comment le future dépend du passé et des événements rares dans le systèmes du vivant. In *La liberté de l'improbable*, ed. Berthoz and Ossola. Collège de France.

Longo, Giuseppe. 2018. Information and causality: mathematical reflections on cancer biology. *Organisms* 2(1): 83–103.

Longo, Giuseppe, and Maël Montévil. 2014. *Perspectives on Organism: Biological Time, Symmetries and Singularities*. Berlin: Springer.

Longo, Giuseppe, Maël Montévil, and Stuart Kauffman. 2012. No entailing laws, but enablement in the evolution of biosphere. In *GECCO Companion '12*. New York: AMC.

Marinucci, Angelo, and Luca Crescenzi. 2021. *Intertestualità. Testo e mondo a partire dalle variazioni*. Pisa: ETS.

Monod, Jacques. 1989. *Le hasard et la nécessité*. Paris: France loisir.

Montévil, Maël. 2022. Historicity at the heart of biology. *Theory in Biosciences* 141: 165–173.

Montévil, Maël, and Matteo Mossio. 2015. Biological organisation as closure of constraints. *Journal of Theoretical Biology* 372: 179–191.

Moreno, Alvaro, and Matteo Mossio. 2015. *Biological Autonomy*. New York: Springer.

Mossio, Matteo, and Alvaro Moreno. 2010. Organisational closure in biological organisms. *History and Philosophy of the Life Sciences* 32: 269–288.

Mossio, Matteo, Maël Montévil, Arnaud Pocheville, et al. 2016. Theoretical principles for biology: variation. In *From Century of the Genome to the Century of the Organism: New Theoretical Approaches*, ed. Ana Soto and Giuseppe Longo, vol. 122, 1–19. Special issue of "Progress in biophysics and molecular biology". Oxford: Pergamon.

Nietzsche, Friedrich. 1999a. Nachlaß 1880–1882. In *Kritische Studienausgabe*, vol. 9. Berlin: de Gruyter.

Nowell, P. C. 1976. The clonal evolution of tumor cell populations. *Science* 194: 123–128.

Rosen, Robert. 2005. *Life Itself: A Comprehensive Inquiry into the Nature, Origin, and Fabrication of Life*. New York: Columbia U.P.

Sonnenschein, Carlos, and Ana Soto. 2007. *The Society of Cells*. New York: Taylor & Francis.

Sonnenschein, Carlos, and Ana Soto. 2016. Carcinogenesis explained within the context of a theory of organism. *Progress in Biophysics and Molecular Biology* 122: 70–76.
Varela, Francisco. 1979. *Principles of Biological Autonomy*. New York: North Holland.
Villoutreix, Paul. 2015. Randomness and variability during embryogenesis A multi-scale approach. Thèse de doctorat. École normale supérieure de Paris.
Weil, Hermann. 1952. *Simmetry*. Princeton: Princeton University Press.
Weinberg, R. 2014. Coming full circle from endless complexity to simplicity and back again. *Cell* 157: 267–271.
West-Eberhard, Mary Jane. 2003. *Developmental Plasticity and Evolution*. New York: Oxford University Press.
West-Eberhard, Mary Jane. 2005. Developmental plasticity and the origin of species differences. *Proceedings of the National Academy of Sciences* 102(1): 6543–6549.

Ontophylogenesis, Interpretation and Symmetries

This chapter proposes the fundamental elements for a relational theory of biological *interpretation*; therefore, it will be outlined how "interpretation" should be understood. Consequently, this chapter will have an epistemological focus.[1]

Since the beginning of this text, an effort has been made to develop an ontophylogenetic discussion, because a theoretical biology has the task of providing an overall picture of life. Obviously, there are processes and elements that refer more properly to ontogenesis and others to phylogenesis, however, the object of theoretical biology (the relational space) can be declined in various forms, from tissue to environment. In this sense, the relationship between contingency, enablement and thickness does not apply only to ontogenetic level, but also to phylogenetic level as already shown in the second part of previous chapter, when it has been stated that biological organizations are historical-empirical *a priori*. The adjectives "historical" and "empirical" allow conceive the evolution within the relational biology. It is possible to *describe* specific real organizations using the *closure of constraints*, but it is *not possible to interpret* their historical evolutions, while biological measurements of relational spaces, acting on enabled possibilities, are characterized by thickness. This approach has an immediate effect on how to conceive historicity of life.

It is certainly possible to conceive biological history or evolution by a pervasive use of causality and from a conservation principle. However, it is possible to oppose to this approach variation, enablement and thickness, in other words, biological contingency. Thus, there are two distinct ways of conceiving history and evolution. In this sense, considering evolution as directed by a fundamental necessity, historicity of life is conceivable either starting from a physical-mathematical causality

---

[1] In order to deepen this way of conceiving "interpretation" from the philosophical point of view, see Marinucci and Crescenzi (2021). A new book about evolution and interpretation is a work in progress.

A. Marinucci, *Theoretical Principles of Relational Biology*,
Human Perspectives in Health Sciences and Technology 6,
https://doi.org/10.1007/978-3-031-39374-7_7

or starting from a principle, whose biological organizations are only particular manifestations.

From this perspective, given an invariant or a general symmetry, what is produced during history is only an expression of it. If *interpretation* has no epistemological value i.e. for Monod, Jacob and for their tradition because everything resolves in a correct copy of code and, therefore, in their "invariant reproduction" (Monod 1989). Similarly, where organization is hired as a principle, one should at least provide a general diagram subsuming all organizations, without falling into physical or chemical explanations: in both cases, an attempt is made to conceive similarity as a degree of equality; on the contrary, it is not the latter that characterizes life, but rather variation.

For this reason, from the philosophical-epistemological point of view, it is necessary to propose new theoretical tools to conceive *contingent historicity* of organisms and their organizations or, more generally, of relational spaces, of theoretical objects of relational biology. This is a central question because if an unity is considered as a principle, then historicity loses its own epistemological status, as it becomes a place of "*revelation*" and not of "*creation*", as Monod himself clearly recognizes (Monod 1989). On the contrary, in relational biology historicity is *creative* precisely because there is no a plan or a general diagram: it is strictly contingent, in the sense expressed in this text.

In fact, if a principle of conservation does not exist and a theory is based on symmetries breakings, then it is necessary to collect historical traces in order to try to build relationships between past and present organisms to reconstruct their evolution. This reconstruction has nothing to do with the *process* time of physical self-organization; on the contrary, it is *historical* time (Longo 2016; Montévil 2022). However, it is not linear because many traces intervene in the formation of different phenotypes and, moreover, the same elements can produce (and have actually produced) different phenotypes or even different forms within the development of same organism (West-Eberhard 2003). In this sense, biological historicity follows symmetries breakings, those bifurcations that occur not during evolution, but *while* evolution happens: there is no organisms evolving *over* time, but the latter is marked by symmetry breakings. Moreover, in the first half of this second part, it has been shown that biological time is an ontogenetic outcome. From the phylogenetic point of view, it is equally a result; it will be shown by interpreting the "density of critical points" (Bailly and Longo 2006), characterizing evolution.

In relational biology, the fact that it is impossible to predict the ear bones from the jaw bones of a Devonian vertebrate (Bailly and Longo 2006) is a clear example of what said. This element is essential because it marks a deep theoretical difference between biology and physics. As seen, it also refers to intrinsic biological indeterminacy and to the fact that biology deals with the constitution of possibilities. In fact, it is a matter of *possible* evolutionary paths that differ from (physical) necessary paths, both towards the future and towards the past. In this sense, it is never possible to establish strictly *causal* links in onto-phylogenetic trajectories. It is about to apply the concepts of enablement and thickness to evolution, also using new philosophical concepts.

The fact that the path leading to the bones of ear is a possible path means that it could have been produced even from something different or that could not have been produced; moreover, from archaeological traces, it is possible to construct *hypotheses* whose connections can *not* explain biological evolution and its trajectories, starting from and using a physical-mathematical necessity.

In other words, it is a matter of dealing with structures of contemporary organisms and the remains of extinct organisms, whose traces can be *interpreted*. From the methodological point of view, according with the idea that biological trajectories are possible, the (onto)phylogenetic paths are constituted while they happen and they are not implementation or, worse still, revelation of an organizational invariant, given *a priori*, neither from ontological nor from structural point of view. Only assuming variation as a principle, history can play an important role within any theoretical biology, since conceiving evolution from "density of critical points" requires to conceive historical-empirical contingency of organisms as irreducible. It is possible to interpret the concept of "*density* of critical points" with the idea that variations mark the time of living beings, when variations happen; strictly speaking, variation does not happen *in* time.

In the previous chapter, it has been shown just how the history of each living being is decisive for measurement of thickness, in this chapter a broader discussion will propose, on the one hand, referring to evolution and, on the other, presenting useful tools to *interpret how* variation occurs. After all, variation and the importance of history are the main theoretical elements that distinguish biology from physics. As shown, relational biology is the *only* approach developing a theoretical thought in which these elements can play a central role. In the sense that they are constitutive of the same biological theoretical object. Indeed, the aim is to develop and to philosophically interpret some central elements of relational biology.

Within evolution various types of organization are not results due to a historical necessity, but they are contingent results, they are historical-empirical *a priori*. In this sense, the *same* initial conditions have produced different results. For example, despite have a common ancestor; human beings represent a different path than to monkey, but both are *realized* possibilities. Now, it is clear that differences are theoretically the most interesting elements, more than the shared elements among living beings. Considering turtles and tortoises, there should exist either just turtles or just tortoises in the same environment, like hurricanes, applying physical necessity and causality rigorously and deeming biological object like the physical one. Nevertheless, starting from the same conditions, the formation of turtles did not prevent the maintenance of tortoises.

However, the question is much more complex:

> Developmental recombination offers an alternative explanation for parallelism in stickleback. The recurrence of this pair of forms suggests that there may be something about the development of their common ancestor that *enables* it to give rise to these two particular forms readily, by means of altered expression of ancestral traits. Research on the ontogeny of anadromous stickleback has revealed that individuals are limnetic when young. They have a slender body form and live in the water column, where they feed on plankton like a limnetic species. Older individuals look and behave like the benthic form, with a stockier

> body and bottom-feeding habits. So the ancestral population occupies both of the habitats observed in the descendent species pairs and exhibits both phenotypes at different times during its life cycle, a pattern that suggests that the different recurrent forms may have originated not by parallel evolution but by altered timing (heterochrony) in the expression of previously evolved adaptive traits. By the heterochrony hypothesis, the limnetic-form species have juvenilized adults, and the benthic-form have full-sized adults like those of the ancestral form (West-Eberhard 2005, p. 6546).

Therefore, the real problem is to conceive philosophically this plurality of evolutionary directions, not only because it is about a real multiplicity, but also because shared or similar traits can be encountered in the most different trajectories. There are countless examples: legs and fins of tortoises and turtles respectively, human arms and front legs of kangaroos etc. Alongside these similar traits, there are "absolute" differences such as Jacobson organ, present in some animals and not in others, echolocation of bats etc. Another philosophically interesting element is that similar or shared traits are maintained, although there are enormous differences among various living beings. Of course, these traits can be bring together starting from its function and from many other reasons; however, it the question of how interpreting them in itself and within an evolution remains open, devoid of necessity and a general conservation principle able to subsume all forms of creation, like a god.[2]

Strictly speaking, the problem is that evolution does not take place over time, but it is precisely the latter that is formed through the biological *density* of critical points. In form of question: how can a historical representation be given, if biological time is a result? Before addressing this issue, it is necessary to answer a preliminary question: how to conceive the *set* of traces that unite and distinguish singular living beings, races, species,[3] etc. within evolution? This chapter attempts to answer these questions. Obviously, these answers are not definitive, but coherent attempts in the horizon of a new biology.

Organisms, Plurality and Interpretation

Before proposing basic elements for a *relational theory of biological interpretation*, it is necessary to clarify beforehand, some concepts that do not come from biology, but that can be applied to biology profitably, *mutatis mutandis*, precisely because the latter claims a contingent historicity of which science has never dealt with.

[2] It is worth remembering that, from the point of view of theoretical construction, it is necessary to overcome the necessity and the idea of existence of a general invariant in order to build a new theoretical biology. From the descriptive point of view, if there are many differences in considering god, organization or DNA as biological invariant, methodologically it is exactly the same because it is a matter of extremely similar theoretical structures.

[3] It is not trivial to remember that these are concepts and not real entities.

The main problem with concepts that have been criticized, abandoned or simply left in background so far is that they mainly refer to a semantic network that has blocked the birth of new ideas. For example, causality limits any kind of theoretical argumentation to physics, to necessity and, as regards the interpretation of traces, to the chain of being and to other secularized theological concepts, such as "revelation" by Monod. On the contrary, symmetry breakings, enablement, thickness, and what marks the theoretical specificity of biology, compared to other scientific disciplines, allow to shed a new light on life. In fact, symmetry breakings and multiple bifurcations allow to conceive evolution of organisms newly, since their contingency gives a new scientific and epistemological status to history.

Before introducing new concepts, it is necessary to understand how mathematical tool of symmetry can be applied to evolution of organisms.

The Myth of "Origin"

From the physical point of view, considering the solution of any system of differentiable integrable equations, particular integral determines a trajectory necessarily. In addition, it is possible to identify an *origin* of movement and their causes. On closer inspection, it is well known that dealing with integrable equations means dealing with a physical epistemic indeterminacy, different biological indeterminacy,[4] that is certainly closer to that of quantum mechanics (Buiatti and Longo 2013). In addition to indeterminacy, there is an essential difference between physical object and the biological one, claiming a central role for historicity. One of the assumptions of applicability of differential equations to physics is that physical object does not change during movement. On the contrary, biological object is characterized by changes, during its onto-phylogenetic evolution.

Looking at history of evolution, it is possible to reconstruct some historical *a priori* (for example, extinct species), but it is not possible to know necessarily their trajectories and their conditions of existence. However, it is possible to *interpret* data in order to reconstruct what certain changes enables, based on specific assumptions. Furthermore, the directions of enabled possibilities can be better characterize through the concept of thickness. Hence, since it is not possible to reconstruct or foresee the appearance of a new character, it is not correct to talk about "origin", because the same initial conditions and the same historical-empirical *a priori* (organization) have produced several different phenotypes. In other words, along evolution all or many of possible paths of various bifurcations, produced by multiple breaks of the same symmetry. Hence, what is really important is not "the origin itself" (assuming that it is completely knowable) as much as produced variations. If two species have a common ancestor, its identification is not sufficient to completely determine the origin of the derived species: the essential element is

[4] For an opposite position see: Huxley (1943) and Grafen (2014).

not so much the common genus, but more the specific difference. Of course, the ancestor or the common trait are important, however it is necessary to conceive them differently. In other words, what needs to be focused on is not the "origin", but the "*point of arising*" ("*punto di insorgenza*"[5]) of variations. Beyond a change of perspective intrinsic to a biology that can conceive contingency and historicity of life, the concept of "origin" is philosophically very compromised with a semantic network, referring to metaphysical and theological approaches. These are linked to a type of principles that have produced the idea of building science based on invariants, conservation principles and symmetries.[6]

It is crucial to talk about points of arising and not about origin because, on a closer inspection, if evolutionary paths, leading from one phenotype to another, are formed during the constitution of trajectories, then they show the place where "*segnature*" (Agamben 2008; Marinucci and Crescenzi 2021) or "traces"[7] of a certain character occur. Strictly speaking, for this reason it is not correct to talk about "origin" as this is *dissolved* in possible (onto)phylogenetic paths and in symmetry breakings. In a phylogenetic point of view this way to thematize the concept of "origin" is due to the specific concepts of "relationship" and "relational space", typical of relational biology. In fact, relations *are* the relational space; its structure, its organization depends on how relations are configured and not on a pre-constituted teleological (and teleonomic) plan.

This allows to conceive the traces of appeared organisms (and not only of common ancestor) as *points of arising* (punti di insorgenza) of living organisms. In other words, the latter contain *traces* of extinct organisms, however, such traces do not present exactly as they are at points of arising, but they are resemantized, they are characterized by variation. This is due to the fact that biological possibilities are formed along the evolution and that evolution is creative. The fact that biology focuses on *possibility* implies two things: the first one is that biological elements (phenotypes) can maintain over time, the second one is that they have always to be *reinterpreted* by evolution and in evolution, as there is no organizational structural invariant.

Symmetry and Plurality

An organism, as well as its specific organization, is a possible result of its contingent history. It represents a group of symmetries, i.e. a set of coherent properties

[5] Although this concept is from Agamben (2008), it is used according to Marinucci and Crescenzi (2021). This concept is important to emphasize that the common ancestor itself is not so important, but what is relevant in it. If the loss of the kiwi's wings is taken into account, its digestive system is an irrelevant element.

[6] In this sense, the concept of "origin" can be considered as the secularization of theological concept of God.

[7] This last word will be preferred. It is from Longo (2016).

qualifying it as a particular relational space. Through symmetries it is possible to conceive an *organism* as a *plurality* of (relationed) elements and not as something unique. It is easy to guess that this allows to conceive *more evolutionary directions starting from the same phenotype*, as they convey different symmetry breakings, starting from the same group of property. From the mathematical point of view, symmetry breakings and subgroups are possible, depending on what properties are preserved or not (Tapp 2012). In this sense, both differences and common or similar traits are theoretically admissible, even in very different phenotypes, but with a common ancestor. Considering humans and apes, contingency has *enabled* new properties in the common ancestor, however, the variation of P_h values with respect to P_p of a given property (or of a set of symmetry of ancestor group) produced different phenotypes, as well as the *maintenance* of a group of specimens and, then, selection acted over all. In this regard, a good example can be the kiwi (West-Eberhard 2003).

This argument goes far beyond the case in which two different phenotypes have the same ancestor. In fact, it also affects the common traits between phenotypes with different ancestors. Depending on the trait chosen for description, similarities and differences can be highlighted not only between apes and humans, but even among monkeys, cows and whales, as they are all mammals. In this case, in different phenotypes, historical contingencies have enabled possibilities in which P_h was such that the same variations occurred, directed by enablement.

In other words, through a single theoretical instrument (theory of symmetries) it is possible to conceive each phenotype as a *coherent plurality* of properties; moreover, by shifting the focus on properties it is possible to conceive the evolution[8] *in different directions*, thanks to enablement and thickness.

Of course, it is necessary to deepen this issue from the philosophical point of view, in order to have tools to interpret and compare traces or specific properties of different phenotypes along the evolution.

It is necessary to talk about "interpretation"[9] because there are no necessary developments of organisms, as biological trajectories are *possible* paths. In cases where a strong physical causality is admitted or where "the program doesn't learn from experience" (Jacob 1970, p. 11), it is not necessary to talk about interpretation, because necessity comes into play. In fact, interpretations should not necessarily be considered mutually exclusive, since the same organism (historical-empirical *a priori*) can take two or more different evolutionary paths, without having to remain faithful to a general plan, a progress or a teleology.

---

[8] In order to avoid to fall into the organicistic perspective, it is important to remember that organisms are historical-empirical *a priori*.

[9] It is important to talk about "interpretation" because this binds the discourse to an immanent and contingent plan as "inter-praetium" means etymologically a negotiation, the fact of establishing the price of something between two or more people. Obviously, for each negotiation different prices are possible; similarly, in relational biology, the same elements can produce different phenotypes.

> As we'll see with biological evolution in section 4.2, novelty is often the result of a reorganization of traces (phenotypic and genetic, as appropiate) from the past. Typically, in order to carry out a future action, it is possible to use and recombine different experiences or (pre-conscious) traces from the past; the use of one or the other, or a mixture of the two, may depend on minor, non-measurable differences in the present - a form of bifurcation thanks to the traces of past histories. Retention and protension are both an interpretative construction and reconstruction of the past (Longo 2016, p. 14).

The specific reference of this quotation concerns retention, memory and protection, but, more generally, as this article shows, it tries to provide a direct explication of what actually happens in "interpretation of traces". However, beyond the necessary position of problem, it is necessary to deepen this issue as it needs general and new philosophical tools, conceivable only in a relational biology. Of course, many studies already exist, but they are unfortunately insufficient because they do not give a real epistemological value to variation, always considered starting from an unity. In other words, based on the elements that can be gathered in evolutionary history of organisms, it is necessary to find philosophical tools able to apply relational biology to evolutionary contingency, without betraying its principles. Several concepts of these have already been proposed, but the concept of "interpretation" has not yet adequately developed.

Traces and Symmetries

Now, it is necessary to characterize more closely the *interpretative act* and to clarify what is meant by "*traces*", "*symmetries*" and "*symmetry breakings*" in this context.

In this type of interpretation, the points of arising (*punti di insorgenza*) of certain characters or properties are not so (or only) important (these are some presuppositions of interpretation), *relationships* between them are more important because they can create new phenotypes and/or organizational typologies. Besides the possibility to explain *emerging* meanings from implementation of relations, relational interpretation allows to consider environmental contingent context as an essential *punto di insorgenza* within the theory. On the contrary, in Jacob (1970) contingency occurs only in the choice of the possibilities and not in their constitution. One of the main elements is precisely the constitution of the context within the theoretical framework and not its generic reference within the interpretative act.

If it is possible to interpret new phenotypes by emphasizing relationships (*even considering the same elements*), then properties of a symmetry group or qualities of hopefulmonster are certainly various, but still linked to the points of arising. Moreover, the fact of considering organism as a relational space and the fact of understanding it mathematically through symmetries makes possible to conceive it as a plurality.

It is necessary to introduce the concept of *trace* (Longo 2016) and better characterize *symmetries* and symmetry *breakings*, precisely in order to concretely characterize relationships giving rise to variations and in order to offer a theoretical idea of "context" that can not more conceived as a vague and elusive notion (and, therefore, difficult to use *rigorously*).

Both from the point of view of interpreter and from the point of view of evolution, in the act of interpreting traces, it is a matter of dealing with points of arising and with resulting phenotype. The sets of properties and symmetry groups are pluralities characterizing *respectively*, on the one hand, a studied phenotype and, on the other hand, each point of arising.

From symmetries of points of arising to that of studied phenotype, there is a set of intermediate, *diachronic and synchronic* steps, in which certain properties are modified so as to approach its final form. This is precisely the terrain on which interpretation works. Hence, starting from traces it is necessary to constitute an interpretative field containing all the relevant traces, *in a given interpretative perspective*, in order to constitute a new interpretation of considered symmetry groups.

Since such a field represents the basis supporting interpretation, it too constitutes a set, in particular, it is the set of properties that *actually* make possible to *conceive* any phenotype. It is a different set from the one inside a specific phenotype, but, since it originates from traces, it is better to clarify, first of all, the latter and then return to symmetries.

Biological Traces

It is necessary to describe separately traces and symmetries and then a more functional characterization. Each point of arising carries traces that may occur in hopefulmonster, in its symmetry group. However, with respect to a given phenotype, there is a series of important *variations recognized*[10] *as relevant* for the reconstruction of final group between the points of arising and hopefulmonster. In fact, it is highly unlikely that a phenotype will form without taking into account changes occurring within the environment and other conditions that enable it.

Before delving into them, it should be noted that the same "point of arising" is such as it is also *recognized as such*; hence, an absolute *origin* is *not* conceivable, but *more and different* points of arising are possible, depending on each interpretative perspective. Therefore, they are reconstructed as such and all the modifications, that reach the final form, are linked to them. This way to talk about points of arising is necessary to reiterate the perspective and the poietic side of every interpretation, as

[10] It is necessary to express in this way because one is always within an interpretation and, therefore, it is not possible to talk about essential causes for the reconstruction of a phenotype, such as Laplace or as in organicist closure of constraints. For the same reason, the points of arising are also related to a specific perspective. Therefore, an element is *essential* for the interpretation of a phenotype within a context and never absolutely.

it is not a matter of "reducing" the final phenotype to its "origins", but of identifying what *enables* it.

Phenotype modifications are normally conveyed by new elements, in the furrow of which the final phenotype is placed, either by a particular reception of the same structural theme in a contingent environment or by an infinity of other means. In this sense, "traces" are all *variations* on a given theme that have been produced *and* that are considered *relevant* for the interpretative *re*construction of final phenotype or of one of its function. Within an interpretative perspective, these variations are what is considered *relevant* in order to understand a phenotype in different points of arising. For example, if an organic or environmental element is a point of arising, it is not necessary to consider it totally for interpretation, precisely because interaction creates new possibilities. Everything that is relevant for interpretation represents a trace, which is such with respect to what emerges from other points of arising. There are various examples: the kiwi, its missing wings, the shape of its beak and the amount of insects on the ground; the bipedal goat etc.

More generally, only considering that interpretation emphasizes *variations*, in the path from the points of arising to hopefulmonster, *traces* can emerge. They allow to reach the constitution of a *coherent* set of differences, whose interaction enables hopefulmonster. The differential element is derived from how points of arising interact in an interpretive act. Already at this point it is clear that, more important than the latter concept is undoubtedly the *set of variations* constituted by traces. As mentioned, such a set constitutes a set of symmetry, a set made of differences.

However, these differences will neither cause nor necessarily imply the final form of a hopefulmonster, but they *enable* it, as these traces are precisely the elements that "deform" interpretative field. As seen, they act *directly* upon the constitution of sense horizon in which hopefulmonster is *enabled* and in which it is *possible* to conceive it. they act indirectly too upon the *real* constitution of hopefulmonster (always contingent). Precisely the way in which traces interact produces *a specific enablement* that directs interpretation towards specific new properties. Since it is about *possible* and unnecessary evolutionary paths, thickness allows to follow evolution of a certain property that can disappear or radically change, at any time. As seen in the previous chapter regarding cancer, if something is possible (enabled), this does not mean that it has to be realized. Evolution has shown that not all tortoises have become turtles. Although the latter were enabled, thickness ("towards") of enablement was such that $P_h \gg P_p$ or $P_h > P_p$ with $P_p \gtrneqqless 0$.

Symmetries and Interpretation

If traces refer to relevant variations enabling a new phenotype, symmetries and symmetry breakings show how those variations are structured. From time to time, each phenotype can have a given interpretation in its interpretative field. It is important to underline that the term "interpretation" always refers to the plurality proper to different phylogenetic trajectories. These may derive from the same ancestor and, therefore, symmetries allows precisely to dissolve each unity, organization and ontogenetic and phylogenetic form into a coherent plurality of elements.

It is possible talk about "symmetry" when an object, subjected to transformations, keeps its *constituent* properties unaltered; on the contrary, there is a break in a symmetry when the latter are modified.[11] Generally speaking, talking about symmetry, the properties form a "group" describing it. Therefore, from the biological point of view, a symmetry expresses a historical-empirical invariance. The group of elements of different points of arising, which constitutes the interpretative field, constitutes a set. Besides being relative to a single interpretation, it is useless to repeat that this type of invariance is an evolutionary *result* and not an essence or a fundamental invariant.

As said, one of the greatest advantages of being able to consider an organism through symmetries is that it immediately becomes something *plural*, a group of properties. Since it is not possible to reach an essence, but only to move from one interpretation to another, every set of properties that enables a hopefulmonster is, therefore, an invariant (*a posteriori*) with respect to *a* specific phenotype.

Obviously, as evolution shows, several interpretations are possible starting from the same phenotype, as its symmetries can be broken variously. After all, a very implementation of interpretation requires precisely a search for new relationships or differential elements in order to be able to interpret differently. It is precisely here that the advantages of conceiving a phenotype or its function as a *plurality* emerge. In fact, as the history of evolution shows that it is *not* necessary that all properties change simultaneously and with respect to the same differentiable elements, in order to produce a variation. In this sense, new symmetries are created, they will be the basis for new interpretations.

From this point of view, the concept of "symmetry breaking" provides theoretical and practical tools to conceive that, with the words of Wittgenstein, "you have to say new things, and yet many old things. [...]. You still have to carry something old. But for a *construction*" (Wittgenstein 2001, p. 82).

[11] The classic example of symmetry breaking is that of the iron-magnetic transition. Considering an iron bar and the same bar, but magnetized, it is easy to understand that its properties change. In particular, the group, constituting the properties of iron bar, differs from that of magnetized bar, even if some properties are maintained, such as rotation on the axis of magnetic field. It is very interesting that symmetries breaking does not necessarily imply the change of all properties, because this aspect allows to free intertextuality from any type of holism.

The Role of Symmetries and Traces

After outlining what is meant by "group", "symmetries" and "traces" in general, it is necessary to clarify their *use*. As mentioned, it is possible to talk about symmetry and group at two levels , within the relational approach:

1. as a coherent set of properties constituting a phenotype or, more generally, a relational space;
2. as the set of traces permitting the interpretation of a phenotype or a part of it.[12]

It is very important to point out that the simply identification of traces is not enough to constitute an interpretative field, but it is necessary to make them interact, compose or, better, *put* them *in relation* because the same elements can produce different effects, interacting differently.

In order to exemplify the first use of symmetries, it is possible to consider the kiwi. Like any other phenotype, it never exists in itself, but it is always found in networks of historical-biological relations, thanks to which it assumes certain characteristics, from time to time. These represents a group of symmetry; in other words, the latter appoints the group of properties or qualities conferring a specific historical meaning to a phenotype. It is certainly trivial to say that the reasons why kiwi has no wings are completely different from those for which a dog does not have wings; it is not trivial to propose a relational model in which these two animals can be interpreted from different traces, although referring to the same property.

Of course, this does not mean that there can be no similarities between different phenotypes, but these become interesting when they show variations, when they show historical singularity of particular organism.[13] As said, it is possible to consider every biological object as *plural* through symmetries, highlighting its variations. Gorilla, humans and chimpanzee represent three specific pluralities, showing three different ways of reorganizing a given group of symmetries, which have given rise to *three different trajectories* from *shared* points of arising.

The second use of symmetry group or property set is a little different. In this case, from the mathematical point of view, it is better to talk about set and not about symmetry group because identity is not an inside property of this group (Tapp 2012). In the interpretative act, a certain property of a phenotype (or limit this as a whole) is interpretable as it is placed within a specific *field of interpretation*, namely a field consisting of that phenotype and its properties. In other words, resuming what already said, the path from the points of arising to the final phenotype is *not* immediate, but requires intermediate steps (traces) composing the set *making* the

[12] From the strictly mathematical point of view, this set can not be considered a "group" because it does not contain its identity, but it is found outside, in the phenotype they explain. For this reason, the more generic expression was used, "set of properties" and not "group". Nevertheless, phenotype is and remains that with respect to which a certain group of properties is organized.

[13] As already mentioned, similarities should not be considered as a degree of identity, but as an evidence of variation. This does not mean that common aspect remains important.

phenotype *conceivable* and possible. Traces are extracted from points of arising because they represent relevant and fundamental variations of meaning (i.e. of form or use) for a given interpretation. Therefore, once relevant traces have been established, interpretation organizes them in order to reconstruct a phenotype. In summary, the use of this set of properties shows all *relevant* traces to a given interpretation.

Symmetry *breakings* play a central role in the process of interpretation, as they represent how, from time to time, the properties of the same phenotype change up to final form. Whether each interpretation modifies or builds a new relationship between the traces that make up the symmetry of interpretative field, it *deforms* the latter in order to open a new possibility on what is interpreted.

It should be reiterated that a "symmetry breaking" does *not* necessarily imply the complete distortion of *all* the properties of a group, but they can be *creatively reorganized* into another group of symmetry. In other words, subjected to the typical transformations of any relocation, if a property group constituting an organism does *not* remain constant, initial symmetry breaks. The symmetry breaking at the point of arising is what actually sets interpretation in motion.

From this perspective, *relation* (constituted by the new properties and by those that have eventually been maintained) constitutes new evolutionary possibilities.

Traces and symmetries play complementary roles within the relational approach. In fact, symmetries describe the process of evolutionary interpretation in its *initial* moments (points of arising), *final* and with respect to the formation of an interpretative field (set of traces). Instead, focusing on traces, the process leading to symmetry breaking of final phenotype is highlighted.

As mentioned, if it is *not* absolutely necessary for *all* properties have to be modified, the extreme cases in which there is no symmetry breakings and in which there is a complete change of properties are of course admissible. It is noteworthy that in the latter case points of arising and their interpretative field always have to be considered, although properties change completely. In other words, no phenotype (or hopefulmonster) completely leaves its context in which it is possible or the interpretative relations that constitute it.

The Interpretative Field

If points of arising *show* partially the perspective from which a phenotype is interpreted, then founding traces means focusing attention on properties that *really* intervene in formation of a phenotype. In other words, traces are what is relevant at points of arising for interpretation. For example, in the case of kiwi, some traces, definitely important for the loss of wings, are its historical contingency, the fact that it is insectivorous, the fact that the environment presents large amount of food on the ground and that there are no terrestrial predators (environmental contingency). Now, other elements certainly come into play, but the point is that someone, among countless environmental and phylogenetic variables, are relevant and, therefore, become "traces". From the *epistemological and theoretical* point of view, these traces do not "explain" the *real* kiwi, but *interpret it as possible*, they constitute its possibility. In other words, starting from traces, kiwi is *not causally determined*, but it is possible to conceive and constitute its *enablement*. It is characterized by thickness, always *within the domain of possibility*. Thickness provides information on the type of contingent relationships, characterizing traces so that a *specific* enabled phenotype develops. Here again, thickness is necessary as relational biology treats causality as described in the previous chapter. This is due to the fact that, as evolution shows, different phenotypes can be produced from the same elements and, moreover, properties of their relations can develop in unexpected forms, precisely because they are possible paths.

Now, by focusing on variations, it is possible to build the *set* of *all traces* affecting *enablement* of kiwi, *always within* a certain interpretative perspective.

Generally speaking, this type of interpretation allows to build a *field* of relations in which different traces form the set that enables it (and does not cause it), with respect to a structural theme or a phenotype. Thus, this set allows to *conceive* eventually real possibilities. Hence, when an element of a phenotype is interpreted (in the domain of reality) something is also affirmed on the its *field* of relations. Characterizing bipedal goat (West-Eberhard 2003) and its muscular structure, quadruped goats and traces enabling bipedal goat and its muscular adaptation are automatically characterized.

Of course, the *field of relations* does not pre-exist to its elements, on the contrary, it is derived *from how* they are actually interconnected. An interpretative perspective extracts *relationally* traces that constitute what enables hopefulmonster from its points of arising. It is possible to call this set "interpretative field". Such relations distort the symmetries of *historical-empirical a priori*. In other words, interpretation always acts on past field and settled phylogenetic histories. In this sense, it is clarified why it is not enough to identify traces, but it is necessary to put them in relation to build a group of symmetry: it is necessary to constitute a field, the ground on which an hopefulmonster can flourish and can be conceived. Deriving from the *relations* between symmetries of points of arising and from what constitutes its set of traces, interpretative field can not be conceived as an *aseptic* "container" including different symmetry breakings of points of arising. On the contrary, by deformations, the *relations* among traces, evolutionary history, different groups of symmetries, *constitute* an interpretative field, only possible in an interpretative *act*. Right here, all the *relevant variations* of points of arising (traces) form *an* interpretative field, starting from a certain perspective. This concept has been emphasized because what is fundamental is not so much the elements, but more what lies *between* them. Therefore, the way in which interpretation *deforms* interpretative space is the key to every new perspective.[14]

---

[14] The concept of "deformation" comes directly from the fact that relational space derives from general relativity, *mutatis mutandis*. Now, if a mass is inserted inside the space, curvature tensor changes and space changes too(relations among masses). In biology, the deformation of

Interpretative field is the *result* of the specific *relations* between different groups of symmetries, from which significant variations emerge. A different importance is assigned to *each* of these, in relation to every interpretative perspective.

Moving from an interpretation to another, from an interpretative field to its new form, it is now possible to understand in what sense the constitution of a field *deforms* a *pre-existing* interpretative space and what constitutes it. In fact, interpretation always begins from a specific phenotype and considers possible variations. In fact, each given interpretative perspective changes more or less radically properties of other perspectives, breaking symmetries in the exposed ways, from which a respective modification of interpretative space will result.

The interpretative field is a way to characterize the observable of relational biology (relational space) from the point of view of *phylogenesis*.

If enablement represents the horizon with respect to which the possibility of hopefulmonster can be interpreted and if thickness allows to consider possibilities as liable to enter into reality, once a variation manifests itself, it can act on a historical *a priori*, fundamentally in two ways. In the first one, an organism can experience variation is by *incorporating* it, without symmetry breakings (subgroup) occurring. This option has already mentioned, also conceivable in organicism, when it was said that, despite variation, thickness does not change and organism becomes stronger ($P_p > P_h \rightarrow P_p \gg P_h$). The second one occurs when, given a variation, there is a symmetry breaking (from $P_p \gg P_h$ to $P_h \gg P_p$). Here, there is a deformation of interpretative field.

Generally speaking, the concepts presented in the previous chapters can be used to explain evolution and to build a general picture of life. Obviously, it is necessary to deepen both the meaning and use of interpretation and the relational biology applied to evolution.

Bibliography

Agamben, Giorgio. 2008. *Signatura rerum*. Torino: Bollati Boringhieri.

Bailly, Francis, and Giuseppe Longo. 2006. *Mathématiques et sciences de la nature*. Paris: Hermann.

Buiatti, Marcello, and Giuseppe Longo. 2013. Randomness and multi-level interactions in biology. *Theory if Biosciences* 132(3): 139–158.

Grafen, Alan. 2014. The formal Darwinism project in outline. *Biology & Philosophy* 29: 155–174.

Huxley, Julian. 1943. *Evolution, the Modern Synthesis*. New York - London: Harper and Brothers Publishers.

Jacob, François. 1970. *La logique du vivant*. Paris: Gallimard.

Longo, Giuseppe. 2016. Comment le future dépend du passé et des événements rares dans le systèmes du vivant. In *La liberté de l'improbable*, ed. Berthoz and Ossola. Collège de France.

---

interpretive field means enabling something having effects on phenotypes. Obviously, it would be necessary to consider the differences between physical and biological relational space. This topic will be dealt with in another text.

Marinucci, Angelo, and Luca Crescenzi. 2021. *Intertestualità. Testo e mondo a partire dalle variazioni*. Pisa: ETS.

Monod, Jacques. 1989. *Le hasard et la nécessité*. Paris: France loisir.

Montévil, Maël. 2022. Historicity at the heart of biology. *Theory in Biosciences* 141: 165–173.

Tapp, Kristopher. 2012. *Simetry*. New York: Springer.

West-Eberhard, Mary Jane. 2003. *Developmental Plasticity and Evolution*. New York: Oxford University Press.

West-Eberhard, Mary Jane. 2005. Developmental plasticity and the origin of species differences. *Proceedings of the National Academy of Sciences* 102(1): 6543–6549.

Wittgenstein, Ludwig. 2001. *Pensieri diversi*. Adelphi: Milano.

Conclusions

The general principles of relational biology have been exposed in this book. In any case, all the theoretical elements that are the basis of any biology, conceived as a discipline, can not be reduced to physics. Obviously, there is still a lot of work to be done, so much so that this book should be considered as the organization of the theoretical space of biology. In this regard, two other texts, dedicated to the evolution and technical uses of relational biology, are in preparation.

In addition to organizing the theoretical space, an attempt has been made to eliminate conceptual confusions and contradictions, arising when the coherence of a theoretical perspective is not adequately explored. Especially, since the early 1980s, new theoretical elements and technical tools have been introduced without changing the theoretical framework of biology which, to a large extent, are still based on ideas that can be traced back to Monod and Jacob or are too dependent on physics. Other attempts are guilty of theoretical inconsistency: the coexistence of a principle of organization or conservation and the principle of variation is simply contradictory. The only two paths that theoretical biology can take are that of invariant and that of variation. Generally speaking, if variation is taken as a principle then the theory has to explain how and why structures are formed and maintained; conversely, if the theoretical principle is a physical invariant such as DNA or organization, then the theory has to explain variation. Assuming both principles at the same time means building a theory that simply explains nothing. From the point of view of relational biology, variation is taken as a principle. Redefining the biological object and its trajectories means no longer being able to treat it as the physical one. In this sense, new concepts, typically biological (enablement, thickness, etc.), have been introduced, which contribute to newly conceive life.

Summarizing, the principles of relational biology are proliferation with variation and motility. Radically developing these principles means to build a general theory of contingency. This kind of theory is in charge of explaining organization and its maintenance. In this sense, the deepening of the concepts of relational space and time has made possible to consider relational space as the theoretical biological

A. Marinucci, *Theoretical Principles of Relational Biology*,
Human Perspectives in Health Sciences and Technology 6,
https://doi.org/10.1007/978-3-031-39374-7_8

object. This general concept makes possible to understand the different relational spaces in biology: tissue, organ, organism, ecosystem, etc.

The construction of a new theoretical biology made it necessary to differentiate it from physics. If relational biology is a theory of contingency, its domain is that of the constitution of the possible. This necessitated the use of the concept of enablement, the reinterpretation of the concept of cause, and, above all, the introduction of the concept of *thickness* (P), which is essential to deal quantitatively with the qualitative concept of enablement. In a forthcoming book, the medical-biological applications of the thickness concept will be dealt with in detail.

Talking about contingency theory allows to consider organisms and biological organizations as historical-empirical *a priori*. Indeed, they appear *historically* and are characterized by contingent ontophylogenetic evolution, which emphasizes the multiplicity of possibilities and all the forms that life can take.

Bibliography

Aspect, Alain, Philippe Grangier, and Roger Gérard. 1982. Experimental realization of the Einstein-Podolsky-Rosen-Bohm Gedankenexperiment: A new violation of Bell's inequalities. *Physical Review Letters* 49:91.

Bertolaso, Marta. 2016. *Philosophy of cancer. A dynamic and relational view*. New York: Springer.

Boltzmann, Ludwig. 2010. Sul rapporto fra il secondo principio della teoria meccanica del calore e il calcolo delle probabilità relativamente alle leggi sull'equilibrio termico. In *Fisica e probabilità*. Milano: Melquíades.

Bottazzini, Umberto. 1981. *Il calcolo sublime: storia dell'analisi matematica da Euler a Weierstrass*. Torino: Bollati Boringhieri.

Boyer, Carl B. 2007. *Storia del calcolo*. Milano: Bruno Mondadori.

Di Paola, Luisa et al. 2012. Protein contact network: An emerging paradigm in chemistry. *Chemical Reviews* 113:1598–1613.

Einstein, Albert, Boris Podolski, and Nathan Rosen. 1935. Can quantum-mechanical description of physical reality be considered complete? *Physical Review* 47:777.

Euler, Leonard. 1960a. Recherches sur la question des inégalités du mouvement de Saturn et de Jupiter. *Opera omnia*. Serie 2. Lipsia: 25.

———. 1960b. Recherches sur le mouvement des corps célestes en général. *Opera omnia* 25. Serie 2. Lipsia.

Everett, Hugh. 1957. "Relative State" formulation of quantum mechanics. *Reviews of Modern Physics* 29:454–462.

Fox Keller, Evelin. 2003. *Le siècle du gène*. Paris: Gallimard.

Fraser, Craig G. 1987. Joseph Louis Lagrange's algebraical vision of the calculus. *Historia Mathematica* 14 (1):38–53.

———. 1989. The calculus as algebraic analysis: some observations on mathematical analysis in the 18th century. *Archive for History of exact Sciences* 39 (4):317–335.

Ghirardi, Gian Carlo. 2009. *Un'occhiata alle carte di Dio*. Milano: Il Saggiatore.

Gould, Stephen Jay. 1996. *Full house*. Traduction française L'éventail du vivant: le mythe du progrès, Éditions du Seuil, 1997. New York: Three Rivers Press.

Guicciardini, Niccolò, and Gianluca Introzzi. 2007. *Fisica quantistica. Una introduzione*. Roma: Carocci.

Kohl, Peter et al. 2010. System biology: An approach. *Clinical Pharmacology and Therapeutics* 88 (1):25.

Lagrange, Joseph-Louis. 1884a. Essai sur le problème des trois corps. In *Œuvres de Lagrange*, Tomo VI. Paris: Gauthier-Villars.

A. Marinucci, *Theoretical Principles of Relational Biology*,
Human Perspectives in Health Sciences and Technology 6,
https://doi.org/10.1007/978-3-031-39374-7

———. 1884b. Leçons sur le calcul des fonctions. In *Œuvres de Lagrange*, Tomo X. Paris: Gauthier-Villars.

Longo, Giuseppe, Paul-Antoine Miquel, et al. 2012. Is information a proper observable for biological organization? *Progress in Biophysics and Molecular Biology* 109:108.

Longo, Giuseppe, and Maël Montévil. 2012a. Randomness increases biological organization: a mathematical understanding of Gould's critique of evolutionary progress. In *Stephen J. Gould Heritage: Nature, History, Society*. Venezia.

———. 2012b. Randomness increases order in biological evolution. In *Computation, physics and beyond*, ed. M. Dinneen, B. Khoussainov, and A. Nies, 289–308. Berlin: Springer.

Longo, Giuseppe, Maël Montévil, Carlos Sonnenschein, et al. 2015, Dec. In search of principles for a theory of organisms. *Journal of Biosciences* 40 (5):955–968.

Longo, Giuseppe, and Lorenzo Seno. 2018. Digital networks, knowledge and political biases in their understanding and use. https://www.di.ens.fr/users/longo/files/longo_senoNetworksEntrop.pdf.

Montévil, Maël. 2019. Measurement in biology is methodized by theory. *Biology & Philosophy* 35:1–25.

Mossio, Matteo, Maël Montévil, and Longo Giuseppe. 2016. Theoretical principles for biology: organization. In *From century of the Genome to the century of the organism: New theoretical approaches*, ed. Ana Soto and Giuseppe Longo, vol. 122. Special issue of "Progress in biophysics and molecular biology", 40–55. Oxford: Pergamon.

Nietzsche, Friedrich. 1999. Nachlaß 1884-1885. In *Kritische Studienausgabe*, vol. 11. Berlin: de Gruyter.

———. 2001. *Al di là del bene e del male*. Milano: Adelphi.

Perutz, Max. 2007. *Le molecole dei viventi*. Roma: Di Renzo.

Poincaré, Henri. 1890. Sur le problème des trois corps et les équations de la dynamique. *Acta mathematica* 13:1–270.

Soto, Ana, Giuseppe Longo, et al. 2016. Toward a theory of organism: three founding principles in search of useful integration. In *From century of the genome to the century of the organism: New theoretical approaches*, ed. Ana Soto and Giuseppe Longo, vol. 122. Special issue of "Progress in biophysics and molecular biology", 100–106. Oxford: Pergamon.

Soto, Ana, Carlos Sonnenschein, and Paul A. Miquel. 2008. On physicalism and downward causation in developmental and cancer biology. *Acta Biotheoretica* 56 (4), 257.

Tabor, Michel. 1989. *Chaos and integrability in nonlinear dynamics*. New York: John Wiley and Sons.

Waff, Craig B. 1976. *Universal gravitation and the motion of the moon's apogee: the establishment and reception of Newton's inverse-square law, 1687-1749*. Tesi di dottorato. The Johns Hopkins University.

———. 1995. Clairaut and the motion of the lunar apse: the inverse-square law undergoes a test. *The General History of Astronomy* 2:35–46.

Wilson, Curtis. 1995. The work of Lagrange in celestial mechanics. *The General History of Astronomy* 2:108–130.

Wittgenstein, Ludwig. 1999. *Libro blu e libro marrone*. Torino: Einaudi.

The manufacturer's authorised representative in the EU is Springer
Nature Customer Service Centre GmbH, Europaplatz 3, 69115 Heidelberg,
Germany. If you have any concerns regarding our products, please
contact ProductSafety@springernature.com

Printed and bound by CPI Group (UK) Ltd, Croydon, CR0 4YY
01/10/2026
02225917-0001